Climate Change: A Reality Check For 2025

Mapping A Path Forward: From Awareness To Action

M.R. Minarsich

Table of Contents

Disclaimer Notice:

Please note the information contained within this document is for educational and entertainment purposes only. All effort has been executed to present accurate, up to date, reliable, complete information. No warranties of any kind are declared or implied. Readers acknowledge that the author is not engaged in the rendering of legal, financial, medical or professional advice. The content within this book has been derived from various sources. Please consult a licensed professional before attempting any techniques outlined in this book.

By reading this document, the reader agrees that under no circumstances is the author responsible for any losses, direct or indirect, that are incurred as a result of the use of the information contained within this document, including, but not limited to, errors, omissions, or inaccuracies.

Introduction

In recent years, the conversation around climate change has transformed from a distant scientific debate into an urgent personal concern. The warming of our planet is no longer just a topic for experts—it's a reality that affects each of us daily. We feel it in the sweltering summer days that seem hotter than ever before, the increasingly frequent severe weather events, and the unsettling news about melting ice caps and rising sea levels. It's clear: climate change is here, and its impacts are profound.

This book is crafted for anyone who finds themselves deeply concerned about these environmental changes and eager to understand more. It's also tailored for educators and students who delve into ecological studies. Together, we will explore the current state of our planet as of 2025, delving into the complex web of changes and challenges brought on by climate shifts. But more importantly, we'll uncover practical, feasible solutions that you, your neighbors, your city, and even the global community can adopt to forge a more sustainable future.

Consider this book your roadmap through the intricate world of climate science, policy, and eco-conscious living. With ten comprehensive chapters, spread over approximately 29,976 words, we have ample room to not only explain the realities of climate change but also offer you tools and strategies for making tangible differences. Our aim is to foster a mindset shift —one where you're not just a passive observer but an active participant in the transformative journey toward sustainability.

To begin with, let's dissect the core aspects of climate change as they stand today. This isn't simply about increasing temperatures; it's about understanding how these temperature shifts ripple out into various facets of our environment and lives. From altering ecosystems and disrupting agricultural cycles to triggering more extreme weather events, the fingerprints of climate change are multifaceted. By grasping these effects holistically, you'll be better equipped to comprehend why immediate, informed action is essential.

Beyond merely identifying problems, this book emphasizes action. What can you do in your daily life to contribute to a solution? Can small changes really make a difference? The answer is a resounding yes. You'll find here a treasury of actionable strategies, from reducing your carbon footprint to advocating for policy changes within your community. We'll cover sustainable living practices, like efficient energy use, responsible consumption, and green transportation options—which collectively paint a powerful picture of what individual and collective action can achieve.

One of the most compelling angles we explore is the integration of indigenous knowledge with contemporary scientific research. Indigenous communities have long been stewards of the land, possessing deep-rooted wisdom about maintaining ecological balance. Their practices hold invaluable lessons for modern conservation efforts. By combining this traditional ecological knowledge with cutting-edge scientific insights, we advocate for a more nuanced and effective approach to tackling climate issues. This synergy of old and new creates a rich tapestry of solutions that respect cultural heritage while embracing innovation.

As we navigate through this manuscript, you'll discover answers to pressing questions about climate resilience. What locations might offer safety and stability in the face of climatic shifts? How can entire communities build resilience against unexpected environmental changes? These discussions are not just theoretical; they're grounded in real-world examples and best practices drawn from around the globe.

In many ways, this book stands on the shoulders of giants, inspired by thought leaders like David Wallace-Wells, Naomi Klein, Robin Wall Kimmerer, and Elizabeth Kolbert. Each author has contributed significantly to raising awareness and driving conversations about our environmental crisis. However, this book aims to carve out its unique space by weaving together the threads of scientific precision, actionable advice, and respect for diverse perspectives.

Why should you read this book? Because understanding the intricacies of climate change is foundational to participating in meaningful change. Because every step towards sustainability, no matter how small, accumulates into significant impact. And because in this era, being informed is not just beneficial—it's imperative. With the right knowledge and tools at your disposal, you can champion environmental stewardship within your sphere of influence, igniting a positive ripple effect.

Our hope is that through engaging storytelling and clear, factual information, you'll feel both educated and empowered. Whether you're an environmentally conscious individual looking to deepen your impact or an academic seeking a comprehensive resource, this book is designed to meet you where you are and guide you forward.

By the end of our journey together, you'll have a robust understanding of the latest scientific findings on climate change, insight into the most significant impacts observed over the last decade, and a collection of best practices for sustainable living. Moreover, you'll gain perspective on where and how to live resiliently in the face of ongoing climate changes.

So, take a deep breath and prepare to embark on an enlightening exploration. The path ahead may be challenging, but it is also filled with opportunities for growth, learning, and impactful action. Welcome to a journey towards understanding our planet and our roles in shaping its sustainable future. Let's dive in and begin crafting a better tomorrow, one informed decision at a time.

Chapter 1

Understanding Climate Change: The 2025 Update

Understanding climate change in 2025 involves looking closely at the patterns and data related to global temperature rise. Recent trends show a persistent increase in global temperatures compared to historical figures, which is essential to grasping ongoing and future climate shifts. By examining the data collected over the past few years, we can observe notable changes in Earth's average surface temperatures and how these align with predictions made by climate models. This alignment provides a clear picture of warming trends and their implications on a global scale.

In this chapter, we will dive into the specifics of recent temperature trends and their impacts on different regions around the world. We will explore phenomena such as Arctic amplification and its effects on the polar regions, as well as the distinct challenges faced by equatorial areas. Additionally, the role of predictive models in forecasting future scenarios based on current emissions trajectories will be discussed. The chapter will also touch upon the importance of public awareness and educational campaigns in driving climate action. Through careful analysis, we aim to provide a comprehensive update on the scientific insights into climate change as of 2025.

Recent Trends in Global Temperature Rise

Understanding the persistent trends in global temperature increases is crucial for grasping the broader dynamics of climate change. As of 2025, the latest recorded temperature data consistently indicate an upward trend compared to historical figures. This persistent increase poses significant challenges, highlighting patterns that are essential for understanding ongoing and future climate shifts.

One notable observation from recent years is that the Earth's average surface temperature continues to rise steadily. Scientists have collected data showing that each successive decade has been warmer than the previous one, with the last few years breaking records for warm temperatures. These records align with predictions made by climate models over the past decades, indicating a clear and persistent trend.

The impact of rising temperatures is not uniform across the globe. The Arctic regions are experiencing disproportionate warming compared to equatorial areas, a phenomenon often referred to as "Arctic amplification." According to Yamanouchi and Takata (2020), the surface air temperature in the Arctic is increasing more than twice as fast as the global average. This rapid warming presents unique adaptation challenges for Arctic ecosystems and communities who are witnessing dramatic changes in their environment.

In contrast, equatorial regions face different challenges. Although temperature increases here are less extreme, they can still lead to severe consequences, such as heatwaves, droughts, and impacts on agriculture. These differences highlight the varied ways in which different parts of the world must adapt to the changing climate. While Arctic regions grapple with melting ice and permafrost, equatorial regions must contend with shifts in precipitation patterns and water availability.

Predictive models play a critical role in forecasting future scenarios based on current emissions trajectories. These models use complex algorithms and a wealth of environmental data to project possible outcomes. Some models predict moderate warming if greenhouse gas emissions are significantly reduced, while others forecast drastic temperature increases if current emission rates continue. A study published in Geophysical Research Letters indicated that even with substantial reductions in emissions, the Arctic could see ice-free summers before 2050 (Simulations Suggest Ice-Free Arctic Summers by 2050, 2020). These predictive models serve as valuable tools for policymakers and scientists, allowing them to prepare for a range of potential futures.

The accuracy of these models varies, with some being more reliable than others. For instance, the Coupled Model Intercomparison Project (CMIP6) provides a broad range of climate projections and is continually refined to improve accuracy. However, uncertainties remain due to the complexity of climate systems and the influence of numerous variables. Thus, while predictive models offer guidance, it's important to recognize their limitations and continue updating them with new data and insights.

Public awareness of temperature changes also plays a significant role in influencing climate action. Surveys and community initiatives reveal a growing recognition of climate change among the general population. Increased public knowledge about temperature trends can drive political will and foster collective action. For example, community-driven efforts to reduce carbon footprints, such as local renewable energy projects or reforestation programs, demonstrate how awareness translates into concrete actions.

Educational campaigns and public outreach are vital components in this process. They help demystify scientific data and make complex topics accessible to everyone. By providing clear information about the impacts of rising temperatures, these campaigns empower individuals and communities to engage in meaningful climate action. Collective efforts at both local and global scales are necessary to address the multifaceted challenges posed by climate change effectively.

The significance of understanding these persistent temperature trends cannot be overstated. They provide critical insights into the mechanics of climate change, helping to clarify the magnitude and urgency of the issue. Persistent warming trends drive home the reality that climate change is not a distant threat but an immediate challenge requiring concerted efforts from all sectors of society.

Furthermore, acknowledging the varied impacts on different regions underscores the need for targeted adaptation strategies.

Policymakers and stakeholders must consider regional specificities when designing and implementing climate policies. For instance, strategies suitable for Arctic regions focusing on preserving ice and protecting wildlife would differ significantly from those needed in equatorial zones where water management and agricultural resilience are paramount.

The Role of Greenhouse Gases and Their Sources

Understanding the role of various greenhouse gases in climate change is crucial to comprehending the larger picture of our planet's evolving climate. Among these gases, carbon dioxide (CO_2), methane (CH_4), and nitrous oxide (N_2O) are the most significant contributors, each with unique characteristics and impacts.

Carbon dioxide, which has a long atmospheric lifetime, is often considered the primary greenhouse gas due to its high volume emitted from human activities. It remains in the atmosphere for centuries, accumulating over time and leading to a more persistent warming effect. This gas is primarily released through the burning of fossil fuels such as coal, oil, and natural gas. Deforestation also plays a significant role in CO2 emissions by reducing the number of trees that can absorb carbon dioxide from the air, thus disrupting the natural carbon cycle.

Methane, though less abundant than carbon dioxide, is significantly more potent, with a global warming potential approximately 25 times greater over a 100-year period. Its sources are varied, including livestock digestion, rice paddies, landfills, and the extraction and transportation of fossil fuels. Methane is also released during the decay of organic matter in low-oxygen environments like wetlands. Although methane has a shorter atmospheric lifetime compared to carbon dioxide, usually around a decade, its powerful immediate impact makes it a critical target for reduction efforts.

Nitrous oxide, another potent greenhouse gas, has a global warming potential about 300 times that of carbon dioxide over a century. It is less prevalent in the atmosphere but originates mainly from agricultural activities, particularly the use of synthetic fertilizers. Nitrous oxide is also emitted during certain industrial processes and fossil fuel combustion. Its long atmospheric lifetime means that once released, it can contribute to warming for over a century.

The cumulative impact of these greenhouse gases on global warming is profound, particularly because their sources are closely tied to essential human activities. Industry, agriculture, and fossil fuel combustion are significant emitters, with developing nations experiencing rapid industrialization and urbanization contributing heavily to the rise in greenhouse gas emissions. As countries grow economically, their energy demands increase, often relying on fossil fuels. Likewise, agricultural practices, crucial for feeding expanding populations, frequently involve activities that release methane and nitrous oxide.

Human activities have not only increased greenhouse gas concentrations but have also disrupted natural cycles that regulate these gases. The natural carbon sinks, such as forests, oceans, and soil, are now under significant strain. For instance, deforestation reduces the Earth's capacity to absorb CO_2, while ocean acidification, caused by excess carbon dioxide dissolving into seawater, affects marine life and ecosystems. These disruptions create feedback loops that can accelerate climate change; for example, the thawing of permafrost releases stored methane, further intensifying global warming.

Addressing these challenges requires comprehensive mitigation efforts encompassing technology, policy, and grassroots initiatives. Current technologies offer several paths toward reducing greenhouse gas emissions. Renewable energy sources, such as wind, solar, and hydroelectric power, provide clean alternatives to fossil fuels. Innovations in energy efficiency can significantly cut down emissions from buildings and transportation. Additionally, advancements in agriculture, such as precision farming and sustainable land management practices, can help reduce methane and nitrous oxide emissions without compromising food security.

Policy measures are equally crucial. International agreements like the Paris Agreement aim to unite countries in the commitment to reducing greenhouse gas emissions. National policies must support these international frameworks by implementing stringent regulations on industrial emissions, promoting renewable energy adoption, and incentivizing carbon capture and storage technologies. Effective policy frameworks can drive innovation and investment in green technologies, ensuring a coordinated approach to tackling climate change.

Grassroots initiatives also play an essential role in mitigating climate change. Community-led projects often focus on local solutions, such as reforestation efforts, sustainable agriculture practices, and awareness campaigns to reduce carbon footprints. These initiatives not only contribute directly to emission reductions but also foster a culture of environmental responsibility and sustainability at the community level. Engaging individuals and communities in climate action can lead to broader societal shifts towards more sustainable lifestyles and consumption patterns, amplifying the impact of mitigation efforts.

Impact on Polar Ice Caps and Sea Level Rise

In recent years, the impacts of climate change on polar ice have become increasingly evident, with significant consequences for global sea levels. Understanding these changes is essential for grasping the broader effects of climate change and formulating effective policy responses. The ongoing loss of ice mass in key polar regions has notable implications for both seasonal changes and long-term stability.

One of the most noticeable effects of climate change on polar regions is the continuous loss of ice mass. Both the Arctic and Antarctic are experiencing reductions in ice cover, with glaciers retreating at alarming rates. This loss influences seasonal rhythms, leading to shorter winters and longer summers. As the ice melts earlier and forms later each year, seasonal patterns that many species rely on for survival are disrupted. These shifts can affect migration, breeding, and feeding cycles of various polar animals.

The melting of polar ice contributes directly to rising sea levels through two primary mechanisms: land ice melting and ocean thermal expansion. When ice sheets and glaciers on land melt, the water flows directly into the oceans, increasing sea levels. For instance, the Greenland and Antarctic ice sheets hold vast amounts of frozen water; their accelerated melting significantly impacts global sea level rise. In contrast, ocean thermal expansion occurs because water expands as it warms. The increase in global temperatures causes seawater to expand and rise, adding another layer of complexity to the issue.

Projections for future sea-level rise vary, but they typically depend on factors such as the rate of ice melt and the degree of thermal expansion. Under high-emissions scenarios, scientists predict more severe outcomes. According to the IPCC, sea levels could rise between 0.29 to 1.1 meters by the year 2100 if greenhouse gas emissions continue unabated (Mimura, 2013). This variability underscores the importance of mitigating climate change to prevent worst-case scenarios.

Polar ice loss also poses serious threats to biodiversity. The ecosystems in these regions are uniquely adapted to extreme conditions, and even small changes can have dramatic effects. For instance, species like polar bears and seals rely on sea ice for hunting and breeding. As the ice diminishes, these animals face habitat loss and food scarcity, pushing them towards endangered status. Similarly, krill populations in the Southern Ocean, which form the foundation of the marine food web, are heavily influenced by sea ice conditions. Changes in krill abundance can ripple up the food chain, affecting fish, penguins, and whales.

Beyond ecological consequences, the socio-economic implications of polar ice loss are profound, particularly for coastal populations. Rising sea levels threaten to inundate low-lying areas, displacing millions of people and destroying infrastructure. Coastal cities like Miami, New York, and Jakarta are already experiencing increased flooding events, with future projections indicating worsening conditions. For instance, case studies highlight that cities such as Dhaka in Bangladesh are extremely vulnerable, where even a small rise in sea level can lead to massive population displacement due to flooding.

The economic cost of these impacts is staggering. Infrastructure damage, loss of property, and the need for relocation pose colossal financial burdens on governments and individuals. Moreover, saltwater intrusion into freshwater systems can compromise drinking water supplies and agricultural productivity, exacerbating food security issues.

Understanding the full extent of these impacts is crucial for informed policy responses. There is often a significant gap between scientific knowledge and policy actions, highlighting the need for better integration of research findings into decision-making processes. Effective adaptation strategies require collaboration across multiple sectors, including government, industry, and communities. For example, policies can incentivize the reduction of greenhouse gas emissions, invest in renewable energy, and enhance coastal defenses to protect against rising seas.

Furthermore, there should be a focus on international cooperation, given that climate change is a global problem requiring collective action. Countries must work together to honor commitments under agreements such as the Paris Accord, aiming to limit global warming and support vulnerable nations in their adaptation efforts. Bridging the gap between science and policy involves not only understanding the data but also communicating it effectively to policymakers and the public.

Final Thoughts

In this chapter, we've delved into the latest scientific insights on climate change as of 2025. We've explored persistent trends in global temperature rise, revealing a clear upward trend that aligns with longstanding predictions from climate models. This consistent warming highlights significant challenges around the globe, especially in regions like the Arctic, which is experiencing rapid warming at more than twice the global average. The varied impacts across different regions underline the need for targeted adaptation strategies.

We also examined how predictive models play a crucial role in forecasting future scenarios based on current emissions trajectories. These models help policymakers and scientists prepare for potential futures, despite inherent uncertainties due to the complexity of climate systems. Additionally, public awareness and collective action are vital components in tackling climate change. Increased understanding and education about these temperature trends can drive political will and community initiatives, emphasizing the importance of both local and global efforts in addressing this pressing issue.

Chapter 2

Natural Habitats

Exploring natural habitats reveals the delicate balance that sustains life on Earth. Climate change, with its multifaceted effects, is increasingly threatening ecosystems and biodiversity. As temperatures rise and weather patterns shift, both flora and fauna are facing unprecedented challenges. Understanding these changes is crucial to appreciating the intricate relationships within our environment and recognizing the urgency of immediate action.

This chapter delves into the impacts of climate change on various ecosystems across the globe. You'll discover how deforestation diminishes species diversity, leaving animals scrambling for limited resources and altering local climates. We will examine ocean acidification's dire consequences for marine life, highlighting the cascading effects on coral reefs and fish populations. Additionally, we'll explore how shifting migratory patterns disrupt breeding habits and food availability for numerous species. Lastly, we'll consider the role of indigenous practices in effective forest management and the importance of inclusive conservation strategies. By the end of this chapter, you'll gain insights into the gravity of these environmental issues and the pathways toward sustainable solutions.

Deforestation and its impact on species diversity

Deforestation significantly reduces the living space available for wildlife, leading to population decline. Forests serve as habitats for innumerable species; when large swathes of these forests are cleared, animals lose their homes and are forced into smaller areas. This crowding can lead to increased competition for the limited resources left, causing food scarcity and heightened stress among animal populations. As animals struggle to adapt to these changes, many species may experience lowered reproductive rates and higher mortality rates. Some species might face extinction if they cannot migrate or adapt to new environments quickly enough.

The removal of trees also alters local climates and affects water cycles. Trees play a crucial role in maintaining the climate of an area by stabilizing temperatures and influencing precipitation patterns. When forests are cleared, the microclimate within and around these areas changes drastically. Without trees to provide shade, temperatures can become more extreme, which can affect wildlife and plant species that have adapted to stable conditions. Moreover, trees are vital for maintaining the water cycle. They absorb rainwater and release it slowly through transpiration, contributing to local rainfall patterns. In the absence of trees, this process is disrupted, leading to irregular rainfall and potentially more severe droughts or floods. These changes disrupt critical ecosystem services such as pollination and water purification.

Another consequence of deforestation is the increased frequency of encounters between humans and wildlife. As humans expand their settlements and agricultural activities into forested areas, animals are often displaced from their natural habitats. This displacement forces wildlife to move closer to human communities, leading to more frequent conflicts. Such encounters can prove dangerous for both parties; animals may attack humans in defense, while humans may resort to harming animals to protect themselves or their property. Additionally, these close encounters can facilitate the transmission of zoonotic diseases, which are diseases that can spread from animals to humans. For instance, deforestation has been linked to outbreaks of diseases like malaria and Ebola, which are believed to spread more easily when human activity intrudes into forest ecosystems.

An effective strategy to mitigate the adverse impacts of deforestation involves integrating indigenous practices in forest management. Indigenous communities have lived in harmony with nature for centuries, maintaining sustainable land-use practices that ensure the continued health of forest ecosystems. These practices often include controlled burning, selective logging, and agroforestry, which can help manage forests sustainably without causing extensive damage. By leveraging the deep knowledge these communities possess, conservation efforts can be more culturally sensitive and effective. Indigenous practices emphasize a holistic approach to land management, considering the long-term health of the ecosystem rather than short-term gains.

Many indigenous communities view the forest not just as a resource but as a living entity that must be respected and protected. This worldview fosters a sense of stewardship and responsibility towards the environment, which can be instrumental in contemporary conservation efforts. For example, in regions like the Amazon rainforest, indigenous peoples have successfully managed vast tracts of land, preserving biodiversity while supporting their livelihoods. Governments and conservation organizations can learn valuable lessons from these traditional practices and incorporate them into broader environmental policies. Recognizing and supporting the rights of indigenous communities to manage their lands can lead to more sustainable outcomes for forest conservation.

Ocean acidification and its effects on marine life

Increased carbon dioxide levels in our atmosphere are leading to significant changes in the world's oceans, a phenomenon known as ocean acidification. This subpoint delves into how this process is impacting marine ecosystems and the broader environment.

When carbon dioxide (CO_2) from the atmosphere is absorbed by ocean water, it reacts with seawater to form carbonic acid. This new compound dissociates into bicarbonate and hydrogen ions. The increase in hydrogen ions causes the pH level of the water to decrease, making it more acidic. This shift in pH challenges the ability of calcifying organisms, such as corals and shellfish, to survive. These creatures rely on calcium carbonate to build their shells and skeletons, but increased acidity reduces the availability of carbonate ions necessary for this process, thus threatening their existence.

Calcifying organisms like corals and shellfish are integral to marine ecosystems. Corals, in particular, are considered the architects of the sea; they build coral reefs that provide habitat, food, and breeding grounds for numerous marine species. When ocean acidity increases, these coral structures become weaker and more susceptible to erosion and breakage. This degradation leads to a decline in reef health and can result in the collapse of entire marine ecosystems dependent on these reefs for survival. For instance, weakened reefs cannot support the rich biodiversity they once did, leading to reduced marine populations and disrupted food chains.

The impact of ocean acidification extends beyond corals and shellfish. Fish, crucial to both marine ecology and human economies, are also affected. Acidified waters can alter the behavior and physiology of fish. Studies have shown that higher acidity can interfere with fish sensory systems, affecting their ability to detect predators, find food, and navigate. Moreover, the stress induced by acidification can impair their reproductive functions, leading to lower birth rates and making fish more vulnerable to diseases. These disruptions can cascade through the food web, ultimately threatening the stability of marine ecosystems and the livelihoods of communities that depend on fishing.

Given these severe consequences, effective mitigation strategies are essential. Establishing marine protected areas (MPAs) is one such strategy. MPAs can buffer ecosystems against the worst impacts of acidification by reducing other stressors like overfishing and habitat destruction. Within these protected zones, marine life can thrive without the added pressures of human interference, promoting resilience against environmental changes. Additionally, MPAs serve as natural laboratories where scientists can study the effects of acidification in relatively undisturbed settings, providing valuable insights for broader ocean management policies.

Another critical measure is implementing stricter emissions controls. Reducing CO_2 emissions at the source is paramount in slowing ocean acidification. Governments and industries must commit to cutting greenhouse gas emissions through various means, including transitioning to renewable energy sources, adopting carbon capture technologies, and enhancing energy efficiency. Policy initiatives like carbon pricing and international agreements, similar to the Paris Agreement, play a vital role in incentivizing emission reductions and fostering global cooperation in tackling climate change.

Furthermore, public awareness and education efforts are essential in driving collective action. By understanding the far-reaching impacts of ocean acidification, individuals and communities can advocate for stronger environmental policies and adopt sustainable practices in their daily lives. Simple actions, such as conserving energy, reducing waste, and supporting eco-friendly products, can collectively contribute to reducing the carbon footprint and easing the burden on our oceans.

Efforts to address ocean acidification also intersect with broader conservation goals. Protecting and restoring coastal ecosystems, such as mangroves, seagrasses, and salt marshes, can enhance the resilience of marine environments. These habitats act as natural carbon sinks, absorbing CO_2 from the atmosphere and mitigating its presence in ocean water. In addition, healthy coastal ecosystems provide critical benefits, from protecting shorelines against erosion to supporting diverse marine life.

Investing in scientific research and monitoring is another key component of an effective response. Continuous observation of ocean chemistry and marine organism health enables scientists to track changes and predict future trends. Advanced modeling techniques can help identify vulnerable regions and species, guiding targeted conservation efforts. Collaborative research across disciplines, including oceanography, marine biology, and climate science, fosters a comprehensive understanding of acidification and informs adaptive management strategies.

By combining local conservation actions with global efforts to reduce CO_2 emissions, we can protect our oceans from the detrimental effects of acidification. The health of marine ecosystems is inextricably linked to the well-being of our planet, and safeguarding them ensures the sustainability of both natural and human communities for generations to come.

Changes in migratory patterns of animals

Climate change significantly affects animal migration patterns, impacting wildlife movements and survival across the globe. As weather patterns shift and temperatures fluctuate, many species are experiencing disruptions in their traditional migratory cycles. This phenomenon can have profound consequences on breeding habits and food availability.

For instance, some birds are arriving earlier to their breeding grounds due to warmer spring temperatures. While this might seem like a minor adjustment, it can lead to mismatches between the bird's arrival and the peak availability of food resources, such as insects or plants needed for feeding their young. Conversely, delayed migrations can also pose a problem, leaving species without enough time to breed successfully before needing to migrate again.

In addition to timing issues, animals are also changing their migratory routes in response to new environmental barriers. For example, melting ice in the Arctic disrupts the traditional pathways of species like the caribou, forcing them to find new routes that may be longer and more hazardous. These new paths often require increased energy expenditure, leading to higher mortality rates as the animals struggle to adapt to unfamiliar terrain and harsher conditions.

Another example is the alteration of migratory routes by birds who now face urban sprawl, climate-induced habitat destruction, and other obstacles. Birds that once flew over undisturbed forests and wetlands might now encounter cities, roads, and agricultural fields, which can not only lengthen their journey but increase their risk of exhaustion and predation.

Given these changes, conservation strategies must evolve to protect affected species. Addressing altered migratory patterns requires international cooperation since many migratory species cross political boundaries. Governments and organizations need to work together to create safe corridors and protected areas that span countries and continents. This coordination ensures that habitats critical to different stages of the migration cycle are preserved and maintained.

One effective approach includes tracking animal movements using GPS technology to better understand their new routes and identify critical stopover sites. By mapping these routes, conservationists can pinpoint where interventions are most needed, such as restoring habitats or reducing human-wildlife conflicts along migration paths.

Specific examples highlight the urgency of adapting conservation efforts. The Monarch butterfly population, for instance, has seen a dramatic decline, partly due to climate change affecting their migratory and breeding patterns. Once, millions of Monarch butterflies would travel from Canada to Mexico each year, but now, changing temperatures and loss of milkweed – their primary food source – have severely impacted their numbers. Conservation initiatives focusing on planting milkweed and protecting critical resting spots along their migration route are essential for their survival.

Similarly, salmon species face significant challenges due to shifting water temperatures and river flow patterns. Salmon rely on cold, clear streams to spawn; however, warming waters and altered river courses caused by climate change and human activity are making it increasingly difficult for them to reach their breeding grounds. Efforts to restore natural river flows, create fish ladders, and remove barriers like dams are crucial steps to assist salmon populations in their migration.

These case studies underscore the necessity for comprehensive conservation plans tailored to the unique needs of migrating species. Without adaptive measures, the survival of many migrating animals remains at risk as their environments continue to change.

Effective conservation also involves public engagement and education. Informing local communities about the importance of preserving migratory routes can lead to increased support for conservation initiatives. Simple actions, such as creating wildlife-friendly gardens, reducing pesticide use, and participating in citizen science projects, can collectively make a significant impact on migratory species.

Final Thoughts

Deforestation and ocean acidification are just two of the many ways climate change is affecting our ecosystems and biodiversity. From disrupting animal migrations to altering local climates, these changes are pushing both flora and fauna towards the brink. Deforestation not only destroys habitats but also forces wildlife into smaller, more competitive spaces, leading to decreased populations and potential extinctions. Ocean acidification, on the other hand, weakens coral reefs and threatens marine life, which in turn disrupts entire food chains. These environmental shifts underscore the urgent need for effective conservation strategies and a collective effort to reduce our carbon footprint.

Reflecting on these challenges, it becomes clear that integrating indigenous practices and establishing marine protected areas could serve as viable solutions. Indigenous communities have long managed their lands sustainably, offering valuable lessons in forest conservation. Similarly, marine protected areas can help buffer ecosystems against the worst impacts of acidification, promoting resilience. Public awareness and education are also critical in driving meaningful change. By taking collective action today, whether through policy changes or everyday sustainable practices, we can work towards preserving our planet's incredible biodiversity for generations to come.

Chapter 3

Extreme Weather Events: A New Normal?

Extreme weather events are becoming a staple of our changing climate, raising questions about how we prepare for and respond to these increasingly common occurrences. Weather phenomena such as hurricanes, tropical storms, heatwaves, wildfires, floods, and droughts not only challenge the resilience of our communities but also demand that we rethink our environmental strategies. The stark rise in the frequency and intensity of these events underscores the need for deeper analysis and consideration.

In this chapter, we delve into various types of extreme weather events shaped by climate change, shedding light on their patterns and impacts. We start with hurricanes and tropical storms, exploring their growing intensity and shifting geographical ranges. Next, we tackle heatwaves and wildfires, examining how these two phenomena are interlinked and exacerbated by rising temperatures. We then turn to flooding and drought patterns, highlighting the pressing need for adaptive water management strategies. By understanding these trends, we can better prepare ourselves and mitigate the adverse effects on both human life and ecosystems.

Hurricanes and Tropical Storms

The increasing intensity and frequency of hurricanes and tropical storms have become a significant concern in recent years, closely linked to climate change. Understanding these changes is crucial not only for scientific knowledge but also for practical implications in preparedness and resilience.

Trends in Hurricane Intensity

In recent decades, there has been a noticeable rise in the number of intense hurricanes. The past ten years have seen an uptick in category 4 and 5 storms, which are characterized by sustained wind speeds of over 130 mph. These powerful storms owe their increased strength to warmer ocean temperatures, a direct consequence of climate change. The science behind this is straightforward: warmer water serves as fuel for hurricanes, allowing them to gather more energy and become more destructive. For instance, Hurricane Irma in 2017 remained a Category 5 storm for three consecutive days, highlighting the trend towards stronger storms. This rise in intensity has profound implications for coastal infrastructure, human safety, and economic stability.

Changing Patterns of Frequency

Not only are hurricanes getting stronger, but their frequency is also evolving. Data indicates that hurricane seasons are lasting longer, starting earlier, and ending later than they traditionally have. Historically, the Atlantic hurricane season runs from June 1 to November 30, but recent trends suggest that significant storms can occur outside this window. This lengthening season means that communities must be prepared for hurricanes for almost half the year, stretching resources and readiness plans thin. Moreover, more frequent storms mean less recovery time between events, compounding the challenges faced by affected areas.

Geographical Shifts in Hurricanes

One of the more surprising and concerning trends is the shift in geographical patterns of hurricanes. Areas that historically have not experienced severe storms are now finding themselves in the path of these natural disasters. For example, regions in the northeastern United States and parts of Europe have started seeing hurricane impacts, posing risks to infrastructure not designed to withstand such forces. This shift can be attributed to changes in atmospheric conditions and sea surface temperatures, which alter the traditional paths that hurricanes take. As a result, cities and towns in these newly affected areas must reevaluate their building codes and emergency response plans to address the unique challenges posed by hurricanes. Guidelines for these new hotspots should include reassessing the structural integrity of buildings, enhancing early warning systems, and developing community-based disaster response strategies.

Impact on Local Ecosystems

Hurricanes don't just affect human populations; they also have significant impacts on local ecosystems. Storm surges and heavy rainfall can cause flooding that disrupts marine and coastal environments. For instance, coral reefs, which provide essential services like coastal protection and fish habitats, can be severely damaged by powerful waves and debris stirred up by hurricanes. Additionally, increased freshwater runoff from heavy rains can alter the salinity levels in coastal waters, affecting species that are sensitive to such changes. Mangrove forests, which act as natural barriers during storms, often suffer extensive damage, reducing their ability to protect inland areas from future hurricanes.

The loss of biodiversity in these ecosystems can have cascading effects, impacting fisheries and human livelihoods.

Furthermore, the influx of pollutants from urban runoff during storms can lead to water quality issues, harming both marine life and human health. Fisheries, a critical source of food and income for many coastal communities, face disruptions as breeding grounds and feeding habitats are altered or destroyed. Efforts to mitigate these impacts should focus on restoring mangroves and coral reefs, implementing sustainable fishing practices, and improving watershed management to reduce pollution runoff.

Lessons Learned from Recent Events

Recent hurricanes have provided valuable lessons in dealing with escalating storms. Hurricane Katrina in 2005, Hurricane Sandy in 2012, and more recently, Hurricane Harvey in 2017, have shown the importance of preparedness and resilient infrastructure. These events highlighted the need for better forecasting models, more robust emergency response plans, and greater investment in resilient infrastructure. Communities have learned that evacuations need to be faster and more efficient, and that communication systems must be improved to ensure that everyone receives timely information.

Additionally, the role of climate adaptation strategies has become clear. Building seawalls, restoring wetlands, and improving building codes to make structures more wind-resistant are all vital steps in mitigating the impact of future hurricanes. Another lesson is the importance of community involvement in disaster preparedness and recovery. Local knowledge and community networks can play a crucial role in ensuring that vulnerable populations receive the help they need quickly.

Moreover, there's a growing recognition of the need for international cooperation. Hurricanes do not respect national borders, and their impacts can be felt globally through disrupted supply chains and economic losses. Sharing best practices, coordinating relief efforts, and investing in global climate resilience initiatives are steps that can help mitigate the global impacts of these increasingly common events.

Heatwaves and Wildfires

Heatwaves are increasingly becoming a hallmark of our changing climate. Recent years have recorded some of the highest temperatures on record globally, signaling a significant departure from historical norms. The frequency and duration of these heatwaves have seen a consistent upward trend, with more days of extreme heat being reported annually. For instance, data reveals that countries such as India, Australia, and parts of Europe have experienced prolonged periods of scorching temperatures, often lasting weeks longer than historical averages. This surge in extreme heat events is intrinsically linked to rising global temperatures, driven primarily by human-induced climate change.

One of the most visible and devastating consequences of heatwaves is their impact on wildfire seasons. Heatwaves create ideal conditions for wildfires to ignite and spread rapidly. The relationship between extreme heat and wildfire activity is stark; higher temperatures dry out vegetation, making it more susceptible to catching fire. In regions like California and Australia, this has translated into extended wildfire seasons that start earlier and end later each year. Statistics from various environmental agencies indicate a marked increase in both the number of fires and the total acreage burned annually. For example, the 2020 wildfire season in California saw over 4 million acres burned, more than double the previous record (MacCarthy et al., 2023). These fires not only devastate natural landscapes but also lead to significant economic losses and pose serious health risks to nearby human populations.

Different regions around the world are uniquely impacted by the dual threat of heatwaves and wildfires. In California, consecutive years of severe heat and drought have turned vast areas of forest and scrubland into tinderboxes. The state's infamous wildfires, like the Mosquito Fire, have razed thousands of hectares, destroyed homes, and displaced numerous communities (MacCarthy et al., 2023). Similarly, in Australia, the "Black Summer" bushfires of 2019-2020 scorched over 18 million hectares, claiming lives and causing widespread destruction. These fires were fueled by the hottest and driest year on record, underscoring the harsh reality of a warming world. Both regions serve as stark examples of the catastrophic consequences of unchecked climate change.

Addressing these issues requires a multifaceted approach, with policy playing a crucial role in mitigation and adaptation. Existing policies aimed at reducing wildfire risks and enhancing community resilience are diverse but often fragmented. For instance, various fire management strategies, including controlled burns and creating defensible spaces around properties, have been implemented with varying degrees of success. However, the scale and intensity of recent wildfires suggest that more comprehensive measures are needed. One critical area of focus should be on enhancing forest management practices to improve resilience against fires. Ending deforestation and curbing the degradation of forests are pivotal steps in this direction.

Community-led initiatives also hold significant potential in improving resilience to these extreme weather events. Local communities can play a proactive role in fire prevention and preparedness through education and awareness campaigns. Encouraging homeowners to adopt fire-resistant building materials and maintain clear zones around their properties can reduce the risk of fire damage. Moreover, community-based fire response teams can provide rapid assistance during emergencies, minimizing the response time and potentially saving lives and property.

On a broader scale, addressing the root causes of climate change is imperative. Mitigating the worst impacts of rising temperatures and extended wildfire seasons will require substantial reductions in greenhouse gas emissions. Governments and policymakers must prioritize transitioning to renewable energy sources and implementing stringent emission reduction targets. Breaking the fire-climate feedback loop necessitates drastic changes across all sectors, from energy production to land use practices.

In addition to policy measures, technological advancements can offer innovative solutions to mitigate the effects of heatwaves and wildfires. Enhanced early warning systems utilizing satellite imagery and predictive modeling can help identify high-risk areas and preemptively deploy resources. Investing in research and development of heat-resilient crops and urban cooling technologies can further bolster efforts to adapt to rising temperatures.

While data alone cannot solve the issue, it plays a vital role in understanding trends and devising targeted responses. Recent data from organizations like Global Forest Watch has been instrumental in tracking fire activity and tree cover loss, helping to identify hotspots and inform policy decisions. Continuous monitoring and analysis of climate patterns and fire incidents will be crucial in adapting strategies and ensuring effective implementation.

Flooding and Drought Patterns

To explore how climate change is causing shifts in flooding and drought patterns, highlighting the urgent necessity for adaptive water management strategies, we must first understand the multifaceted ways our planet's climate is transforming weather extremes.

Flooding frequency and intensity have shown a marked increase in recent decades. Various regions experience more frequent and severe flooding, largely driven by changes in rainfall patterns. Extreme precipitation events are becoming more common, leading to flash floods that overwhelm infrastructures and natural landscapes. This uptick in heavy rainfall can be linked to the warmer atmosphere's ability to hold more moisture, resulting in intense downpours when conditions trigger rain (Jan et al., 2022). Such sudden influxes of water not only cause immediate damage to property and ecosystems but also have lasting impacts on communities—

destroying homes, displacing families, and straining emergency services.

Examining records over recent years, prolonged inundations are another concerning trend. Areas that once experienced seasonal flooding are now facing extended periods of high water levels, challenging localities unprepared for such persistent conditions. For instance, towns along riverbanks and coastal areas are seeing their critical infrastructures, such as roads, bridges, and sewage systems, frequently overwhelmed by floodwaters, which can disrupt daily life and economic activities for weeks or even months.

On the flip side, drought occurrences are becoming increasingly severe and prolonged across many global regions. As temperatures rise, evaporation rates intensify, and soil moisture diminishes more rapidly during dry spells. This drying trend is starkly visible in places like the American West and parts of Australia, where historical analysis shows a clear increase in the duration and severity of droughts over the past century. Dry soils and dwindling water reservoirs significantly affect agriculture, drinking water supplies, and the overall ecosystem health, leading to adverse outcomes such as crop failures, increased incidence of wildfires, and biodiversity loss.

Understanding these shifting patterns necessitates the implementation of innovative water management strategies. Adaptive approaches are essential to efficiently manage water resources amidst these unpredictable conditions. One promising technique is Managed Aquifer Recharge (MAR), where excess surface water is deliberately redirected into aquifers for storage during times of plenty. This method ensures a reliable water supply during dry periods, providing a buffer against drought impacts. MAR systems often involve infrastructure like percolation basins and injection wells, designed to enhance groundwater replenishment when surface waters are abundant.

Additionally, deploying water-efficient technologies in agriculture can profoundly improve irrigation practices. Techniques such as drip irrigation and the use of greenhouses help conservation by targeting water directly to crops, minimizing loss through evaporation and runoff. These methods not only bolster irrigation reliability under severe scenarios like RCP 8.5 but also support sustainable agricultural productivity.

Community resilience and proactive planning play pivotal roles in adapting to these extreme weather patterns. Local adaptation strategies should involve community engagement to ensure that plans reflect the needs and capacities of the people most affected. For example, developing and implementing adaptive stormwater management practices, such as removing impervious surfaces and replacing undersized culverts, can drastically reduce flood risks. Communities with active involvement in these efforts often see better outcomes, as evidenced by examples from Camden, New Jersey, and Washington D.C., where green infrastructure projects have successfully managed stormwater.

Moreover, restoring and maintaining wetlands is a crucial measure to mitigate flooding. Wetlands act as natural sponges, absorbing excess rainfall and reducing downstream flood risks. Conservation projects aimed at preserving these vital ecosystems should be prioritized. Observational data reveal that well-maintained wetlands significantly dampen the impact of heavy rains and provide valuable wildlife habitats.

In drought-prone areas, integrated water resource management strategies are equally vital. Implementing water storage solutions like reservoirs and ponds creates reserves that can be tapped during dry spells. Furthermore, community-driven initiatives, such as establishing local water conservation groups or promoting drought-resistant landscaping, can supplement larger-scale efforts.

Pilot projects and collaborations among various agencies offer a practical testing ground for innovative solutions. These projects can explore novel techniques and facilitate knowledge sharing across different regions. For example, the Blue Plains Wastewater Facility in Washington D.C. has conducted studies that reinforce infrastructure against floods while simultaneously enhancing operational capabilities through improved monitoring systems.

Planning and insurance mechanisms provide an additional layer of security, ensuring financial sustainability despite extreme weather events. Adequate insurance coverage helps utilities cover the substantial costs associated with disaster recovery and infrastructure rebuilding. Training staff on climate change impacts and adaptation strategies further bolsters preparedness, enabling quicker, more effective responses when crises arise.

Adopting alternative power supplies for water utilities enhances resilience, especially during emergencies. Off-grid energy sources like solar and wind provide dependable electricity, crucial for maintaining water operations when natural disasters disrupt traditional power grids. Relocating facilities to higher elevations can also avert coastal flooding risks, safeguarding critical infrastructure.

Final Thoughts

As we wrap up our exploration of extreme weather events, it's clear that climate change is dramatically altering the landscape of hurricanes and tropical storms. We've delved into how rising ocean temperatures are fueling more intense hurricanes, leading to greater destruction and challenges for coastal communities. Additionally, we observed how the frequency of these storms is extending beyond traditional seasons, forcing regions to adapt their preparedness strategies. These shifts not only reshape human habitats but also stress local ecosystems, such as coral reefs and mangrove forests, which play crucial roles in coastal protection.

Moreover, the geographical patterns of hurricanes are no longer predictable, bringing new hurdles to areas previously unaccustomed to such severe weather. This requires a reassessment of building codes and emergency plans to enhance resilience. The impact on both human life and natural environments underscores the urgency for robust climate adaptation measures. Learning from recent hurricanes, the chapter emphasizes the need for improved forecasting, resilient infrastructure, and community involvement. By fostering international cooperation and proactive approaches, we can better navigate the complex challenges presented by our changing climate.

Chapter 4

Human Health and Climate Change: Emerging Concerns

Human health and climate change are deeply interconnected in more ways than we often realize. As our planet continues to warm, various adverse effects on human health emerge, some of which have direct consequences while others manifest indirectly. Rising temperatures and changing ecosystems create ideal conditions for disease-carrying vectors like mosquitoes and ticks to thrive, thereby facilitating the resurgence and spread of diseases such as Zika virus and malaria. Furthermore, altered rainfall patterns disrupt the natural life cycles of these vectors, promoting unpredictable outbreaks and increasing the global health burden. This chapter aims to shed light on the critical intersection between climate change and public health, offering insight into how warming climates are reshaping the landscape of human well-being.

In this chapter, readers will explore several pressing health concerns related to climate change, starting with the proliferation of vector-borne diseases. We delve into how warmer temperatures and erratic rainfall contribute to the rise of diseases carried by mosquitoes and ticks. Additionally, we examine the heightened risks posed by extreme heat events, which not only increase the incidence of heat-related illnesses but also exacerbate existing health conditions. Mental health impacts are another significant focus, as we discuss the emotional toll of environmental changes and forced migration due to climate events. Finally, the chapter outlines the necessity of a coordinated global response, integrated public health strategies, and robust preventative measures to mitigate these emerging health threats. Through this comprehensive examination, we aim to enhance understanding and drive action toward safeguarding public health in the face of climate change.

Spread of Vector-Borne Diseases

Climate change has far-reaching consequences on public health, one of which is the increasing spread of vector-borne diseases. Warmer temperatures and changing ecosystems provide favorable conditions for vectors such as mosquitoes and ticks to thrive, facilitating the emergence and resurgence of diseases like Zika and malaria. These diseases are carried by vectors that flourish in warmer climates, leading to an increase in transmission rates. For instance, Aedes mosquitoes, responsible for spreading Zika virus, are now found in regions previously unsuitable for their breeding due to rising temperatures (Mojahed et al., 2022).

Altered rainfall patterns further exacerbate the problem. Mosquitoes rely on water bodies for breeding, and changes in precipitation can disrupt their life cycles, leading to unpredictable outbreaks. In areas where rainfall becomes more erratic, periods of heavy rain can create numerous breeding sites, followed by dry spells that accelerate mosquito development. For example, Anopheles mosquitoes, carriers of malaria, find temporary water bodies created by rainfall particularly suitable for laying eggs. Changes in these patterns have led to increased malaria outbreaks in regions like the Ethiopian highlands, where a temperature increase of 0.2°C per decade has exposed nonimmune populations to higher risks (Thomson & Stanberry, 2022).

Increased travel and globalization also intensify the risk of disease outbreaks across borders, turning local health issues into global concerns. People traveling from endemic regions can inadvertently carry pathogens to new areas, sparking outbreaks in places previously free from specific vector-borne diseases. The dengue fever outbreak in France and Croatia in 2010 is a stark reminder of how quickly diseases can spread under favorable climatic conditions (Thomson & Stanberry, 2022). The interconnectedness of today's world requires robust surveillance systems to detect and respond to emerging threats promptly.

Public awareness campaigns and innovations in vaccine development are crucial in combating the rise of these diseases. Educating communities about prevention measures, such as eliminating standing water sources, using insect repellents, and installing mosquito nets, can significantly reduce transmission rates. Moreover, advancements in vaccine technology hold promise for long-term solutions. For example, efforts to develop vaccines against Zika and malaria are ongoing, with some promising candidates undergoing clinical trials. Increased funding and support for such research can help mitigate the impact of these diseases on vulnerable populations.

Preventative measures should be a key focus to curb the spread of vector-borne diseases in the face of climate change. Governments and health organizations must prioritize initiatives that address both immediate and long-term challenges. Implementation of effective vector control strategies, such as insecticide-treated bed nets and environmental management to reduce breeding sites, is essential. Additionally, programs that promote community engagement and education can empower individuals to take proactive steps in protecting themselves and their communities.

A coordinated global response is equally important. Collaborative research initiatives that share findings on vector behavior, climate impacts, and effective interventions can inform public health strategies worldwide. The World Health Organization's Global Vector Control Response (GVCR) framework emphasizes the need for integrated approaches to vector management, including strengthening health systems, improving disease surveillance, and fostering innovation. By synchronizing health planning with climate change strategies, countries can enhance their resilience to vector-borne diseases and safeguard public health.

Furthermore, it is critical to integrate climate data into public health strategies to anticipate and manage health crises effectively. Utilizing climate models to predict vector distribution and disease outbreaks can guide resource allocation and preparedness efforts. For instance, early warning systems based on climate and vector data can alert health authorities to potential outbreaks, enabling timely interventions to prevent widespread transmission.

Developing policies that support sustainable health systems in at-risk areas is another vital aspect of the global response. Investing in healthcare infrastructure, training healthcare workers, and ensuring access to essential medical supplies can improve the overall capacity to manage vector-borne diseases. Additionally, addressing socioeconomic factors that contribute to vulnerability, such as poverty and lack of access to clean water and sanitation, can reduce the burden of these diseases on affected communities.

Heat-Related Illnesses

As global temperatures continue to climb, the incidence and severity of heat-related illnesses become an ever-growing concern. Heatwaves, which are periods of abnormally high temperatures often lasting several days, are increasingly frequent and intense due to climate change. These extreme heat events pose significant risks to human health, making it imperative to examine their causes and consequences.

Rising global temperatures result in more frequent and severe heatwaves around the world. Climate change acts as a catalyst, intensifying these heat events and extending their duration. Cities, particularly those with dense populations and limited green spaces, experience what is known as the urban heat island effect. This phenomenon occurs when natural landscapes are replaced with concrete and asphalt, which absorb and retain heat, leading to significantly higher temperatures in urban areas compared to their rural counterparts. The urban heat island effect exacerbates the impact of already rising temperatures, creating dangerous conditions for residents.

Vulnerable populations bear the brunt of these extreme heat events. The elderly, for instance, are at heightened risk due to their decreased ability to regulate body temperature effectively.

As people age, their thermoregulatory systems become less efficient, making them more susceptible to heat stress and related illnesses such as heat exhaustion and heatstroke. Similarly, individuals with pre-existing health conditions, such as cardiovascular or respiratory diseases, may find their symptoms exacerbated in hot weather. These individuals often require special attention and care during heatwaves to prevent serious health complications.

The homeless population also faces increased vulnerability during extreme heat events. Without access to air conditioning or adequate shelter, homeless individuals are exposed to prolonged periods of high temperatures, putting them at significant risk for dehydration, heat exhaustion, and heatstroke. Their limited resources make it difficult for them to stay cool and hydrated, further compounding their susceptibility to heat-related illnesses. Community-based interventions, such as outreach programs and mobile cooling units, can help mitigate these risks by providing essential services and support to the homeless population during heatwaves.

Chronic medical conditions can deteriorate with sustained exposure to high temperatures. Conditions like diabtes, heart disease, and chronic obstructive pulmonary disease (COPD) are particularly susceptible to worsening in extreme heat. For instance, high temperatures can lead to dehydration, which in turn can cause complications for diabetics who need to maintain strict control over their blood sugar levels. Similarly, individuals with heart disease may experience increased strain on their cardiovascular system, potentially resulting in heart attacks or strokes. It is crucial for healthcare providers to monitor patients with chronic conditions closely during heatwaves and provide guidance on how to manage their health effectively.

Effective mitigation strategies are essential in addressing the health impacts of extreme heat. Community cooling centers serve as vital refuges during heatwaves, offering air-conditioned spaces where individuals can seek respite from the scorching temperatures. These centers are particularly important for vulnerable populations who may not have access to adequate cooling at home. Public awareness campaigns can help inform people about the availability and locations of these cooling centers, encouraging them to take advantage of the resources provided.

Urban planning also plays a critical role in mitigating the effects of extreme heat. Increasing green spaces within cities is one effective strategy. Trees and vegetation provide shade and reduce the ambient temperature through a process called evapotranspiration, where water is evaporated from plant surfaces, cooling the surrounding air. Parks, gardens, and green roofs can help counteract the urban heat island effect by lowering surface temperatures and providing cooler environments for city dwellers. Incorporating green infrastructure into urban design not only helps combat heat but also improves overall air quality and enhances the aesthetic appeal of urban spaces.

Moreover, urban planners can adopt other strategies such as using reflective materials for roads and buildings to reduce heat absorption. Implementing policies that encourage the use of cool roofs, which reflect more sunlight and absorb less heat, can significantly lower indoor temperatures and reduce the overall heat burden on communities. Additionally, promoting the use of public transportation and active modes of travel, like cycling and walking, can decrease the reliance on heat-generating vehicles, contributing to a cooler urban environment.

Policymakers must also consider the long-term implications of climate change on public health and develop comprehensive plans to address heat-related health concerns. Investing in infrastructure improvements, such as expanding access to cooling technologies and enhancing emergency response capabilities, is crucial for building resilience in the face of increasing temperatures. Ensuring equitable access to resources and support systems is essential to protect the most vulnerable members of society from the adverse effects of extreme heat.

Education and community engagement are key components of effective heat mitigation strategies. Public health campaigns can raise awareness about the dangers of heat-related illnesses and provide practical tips for staying safe during heatwaves. Simple measures, such as drinking plenty of water, wearing lightweight and light-colored clothing, and avoiding strenuous activities during peak heat hours, can significantly reduce the risk of heat stress. By empowering individuals with knowledge and resources, communities can become more resilient and better prepared to handle the challenges posed by rising temperatures.

Mental Health Impacts

In the context of human health and climate change, understanding the psychological impacts is crucial. The emotional toll of an ecological crisis often goes unnoticed but can be profoundly significant. This subpoint dives into these psychological ramifications, highlighting why it is essential to maintain mental health amidst such global challenges.

Anxiety disorder related to climate change awareness is on the rise, particularly among younger generations. Known as "eco-anxiety," this phenomenon reflects the deep-seated fear and helplessness individuals feel regarding the future of the planet. The overwhelming influx of alarming environmental news contributes significantly to this anxiety, making younger people worry about their future and the viability of life on Earth. Studies indicate that constant exposure to information about climate disasters, species extinction, and environmental degradation can lead to chronic stress and anxiety, impacting overall well-being.

Forced migration due to climate events is another critical issue leading to severe emotional and psychological challenges. When natural disasters such as hurricanes, floods, or prolonged droughts force communities to leave their homes, the resulting displacement brings about a host of mental health issues. Migrants often experience grief, loss, and a sense of disconnection from their previous lives. The concept of solastalgia, which describes the distress caused by environmental changes close to one's home, encapsulates the trauma experienced by those who witness their environments' destruction, leading to significant emotional turmoil (Bradley Patrick White et al., 2023). These migrants also face acculturation stress as they adapt to new cultures and attempt to rebuild their lives in unfamiliar territories. This dual burden of physical displacement and emotional upheaval can induce depression, anxiety, and other psychological disorders (Padhy et al., 2015).

Developing community-based support networks is a vital strategy for building psychological resilience in areas affected by climate change. These networks provide emotional support and practical assistance, fostering a sense of belonging and reducing feelings of isolation among community members. Local initiatives like peer-support groups and communal activities can significantly mitigate the mental health impact of climate events. For example, after natural disasters, communities that engage in collective recovery efforts often show higher levels of psychological resilience compared to those that do not. Community gardens, local environmental projects, and social gatherings centered around climate action can offer a constructive outlet for eco-anxiety while strengthening social bonds.

Educational initiatives are equally important in preparing communities for the mental health implications of climate impacts. By raising awareness and educating individuals about the potential psychological effects of climate change, we empower them to cope better with these challenges. Schools and universities can integrate mental health education related to climate change into their curricula, ensuring that students understand both the scientific and emotional aspects of environmental issues. Workshops and seminars led by mental health professionals can equip individuals with coping strategies and stress management techniques tailored to climate-related anxiety. Furthermore, public campaigns and information sessions can reach broader audiences, helping to normalize discussions around mental health and climate change and encouraging proactive behavior.

These educational efforts should also promote sustainable living practices as a way to reduce eco-anxiety. Encouraging actions such as reducing waste, supporting renewable energy, and participating in local conservation projects can give individuals a sense of control and purpose, mitigating feelings of helplessness. When people see themselves contributing to solutions, even in small ways, it can significantly improve their mental health. Practical steps towards sustainability not only benefit the environment but also provide psychological relief by turning anxiety into actionable and positive behavior.

Healthcare professionals play a crucial role in addressing the psychological impacts of climate change. Incorporating trauma-informed care approaches, which focus on recognizing and responding to signs of trauma, can be particularly effective in treating individuals affected by climate-related events. Training for healthcare providers should include specific modules on the mental health consequences of climate change, enabling them to offer more nuanced and effective care. Additionally, expanding disaster recovery programs to include long-term mental health support is essential. Immediate emergency responses often overlook the extended psychological recovery process, which can last months or even years after the initial event.

Policies and funding dedicated to mental health services in climate-affected regions can make a substantial difference. Governments and international organizations need to prioritize mental health in their climate action plans. Allocating resources for mental health infrastructure, training, and research will ensure that communities are better equipped to handle the psychological fallout of climate change. Collaborative efforts between policymakers, mental health experts, and environmental scientists can create comprehensive strategies that address both the physical and mental health impacts of climate crises.

Moreover, digital platforms and technology can offer innovative solutions for mental health support. Online counseling, virtual support groups, and mobile apps designed to manage stress and anxiety can reach individuals in remote or underserved areas. These tools can provide continuous support and resources, complementing traditional face-to-face therapy. Integrating mental health services into existing climate change apps and websites can also raise awareness and provide immediate assistance to those in need.

Global Response and Preventative Measures

International cooperation and preventative measures are crucial to addressing the health threats posed by climate change. As climate change continues to impact global communities, collective efforts are becoming increasingly important to safeguard public health and enhance resilience against its negative consequences.

Collaborative research initiatives are a cornerstone of these cooperative efforts. By sharing findings across borders, nations can develop proactive public health measures tailored to specific climate-induced health threats. For instance, research into vector-borne diseases like malaria and dengue fever can inform strategies for mosquito control and early warning systems in regions most at risk. Such collaboration not only broadens the knowledge base but also accelerates the implementation of effective interventions, minimizing the time lag between discovery and action.

Funding is another critical component in building resilient health systems, particularly in at-risk areas. Sustainable financing ensures that health infrastructure can withstand and respond to the pressures exerted by climate change. Developing countries often bear the brunt of climate-related health issues due to their limited resources. International funding mechanisms, such as the Global Fund, play a vital role in bridging this gap, enabling these nations to strengthen their health systems and improve access to essential services. For example, investments in water and sanitation facilities can prevent the spread of waterborne diseases exacerbated by increased flooding or drought conditions.

The integration of climate data into public health strategies is essential for anticipating and managing health crises. Climate models and forecasts provide valuable insights into potential health risks, allowing policymakers and health professionals to prepare and respond effectively. For instance, incorporating temperature and humidity projections into heatwave response plans can help mitigate the impact on vulnerable populations, such as the elderly and those with chronic health conditions. This predictive approach enhances the ability to deploy resources efficiently, ensuring timely intervention and reducing the overall burden on healthcare systems.

Policies that synchronize health planning with climate change strategies are urgently needed. These policies should address both mitigation and adaptation, recognizing the interconnectedness of environmental and health outcomes. For example, urban planning policies that promote green spaces can reduce urban heat islands and improve air quality, subsequently lowering the incidence of respiratory diseases. Similarly, agricultural policies that support sustainable practices can mitigate food insecurity and malnutrition caused by climate-induced crop failures.

A key aspect of international cooperation is the establishment of partnerships across different sectors and levels of government. Effective governance structures facilitate coordination among national, regional, and local authorities, ensuring a cohesive response to health threats. For example, the European Union's Climate-ADAPT platform provides a framework for member states to share information and best practices, fostering a collaborative approach to climate adaptation in the health sector. Additionally, partnerships with non-governmental organizations (NGOs) and the private sector can leverage expertise and resources, enhancing the capacity to implement health initiatives on the ground.

Public awareness and education are also vital components of preventative measures. Informing communities about the health impacts of climate change and promoting adaptive behaviors can significantly reduce vulnerability. Educational campaigns can emphasize the importance of preventive actions, such as using insect repellent in areas prone to vector-borne diseases or staying hydrated during heatwaves. Engaging influential figures and leveraging social media can amplify these messages, reaching a broader audience and fostering a culture of preparedness.

Furthermore, equity considerations must be central to these efforts. The most vulnerable populations, including low-income communities and marginalized groups, often face the greatest health risks from climate change. Ensuring that mitigation and adaptation strategies prioritize these groups is essential for achieving health equity. For instance, targeted interventions in informal settlements can improve living conditions and reduce exposure to climate hazards, while inclusive policies can enhance access to healthcare and social support services.

The World Health Organization (WHO) has highlighted the importance of integrating health into climate change policies. Their response centers around promoting actions that both reduce carbon emissions and improve health, building climate-resilient health systems, and protecting populations from climate-related health impacts (World Health Organization, 2023). This comprehensive approach underscores the need for a multi-faceted strategy that addresses the root causes of health vulnerabilities while enhancing the capacity to cope with ongoing and future challenges.

Final Thoughts

Climate change profoundly impacts our health in ways we may not always see immediately. This chapter has walked us through the spread of vector-borne diseases like malaria and Zika, which are becoming more common as temperatures rise and ecosystems shift. It also highlighted the increase in heat-related illnesses, particularly stressing how vulnerable groups, such as the elderly and those with pre-existing conditions, suffer the most. Lastly, we delved into the mental health consequences, discussing eco-anxiety and the emotional toll of forced migration due to climate-related disasters.

Understanding these various health impacts allows us to take informed actions. From improving public awareness and education to enhancing healthcare infrastructure and policies that integrate climate data, many steps can help mitigate these effects. Developing community support networks and promoting sustainable living practices are essential strategies. By fostering global cooperation and prioritizing equitable solutions, we can better prepare ourselves and protect vulnerable populations from the ongoing and escalating challenges posed by climate change.

Chapter 5

Sustainable Living: Individual Actions with Global Impact

Living sustainably is not just about making grand gestures; it's about the small, everyday choices that collectively make a substantial difference. This chapter explores how individual actions can significantly impact global environmental efforts. Simple habits such as reducing meat consumption, minimizing single-use plastics, embracing minimalism, and practicing conscious consumerism are all within our grasp and can lead to meaningful change. By adjusting our routines and being mindful of our consumption patterns, we can all contribute to a healthier planet.

In the following pages, we'll delve into practical solutions and habits you can adopt to live more sustainably. We'll discuss the benefits of a plant-based diet, strategies for reducing plastic use, the principles of minimalism, and the importance of supporting eco-friendly brands.

Each section provides actionable steps and insights, showing how even minor adjustments in your daily life can help combat climate change. Whether you're looking to make a big impact or start with small changes, this chapter offers guidance and inspiration for anyone committed to living sustainably.

Reducing Carbon Footprints Through Lifestyle Changes

Adopting small lifestyle changes can have a significant impact on reducing carbon emissions, and these changes are within reach for most individuals. One of the easiest steps to take is reducing meat consumption. Embracing a plant-based diet can significantly lower methane emissions from livestock, one of the potent greenhouse gases contributing to climate change. For those who find it challenging to transition completely, starting with initiatives like "Meatless Mondays" or integrating more plant-based meals into daily diets is a manageable way to make a difference.

According to research, plant-based diets not only lower methane emissions but also utilize more sustainable farming methods that require fewer resources such as water and land (Jeni, 2024).

Another impactful change is minimizing the use of single-use plastics. Everyday items like plastic bags, utensils, and bottles contribute largely to ocean pollution and environmental degradation. Revolving around the principle of reuse, switching to reusable products can substantially reduce waste. Simple actions like carrying a reusable shopping bag, water bottle, and coffee cup can cut down on single-use plastic usage effectively. This shift also encourages a broader culture of eco-friendly practices within communities, promoting sustainability at a larger scale.

Adopting minimalism is another powerful approach to sustainable living. Minimalism involves focusing on owning fewer possessions and prioritizing quality experiences over material goods. By consuming less, individuals can significantly reduce their carbon footprint. This lifestyle not only cuts down on waste but also prompts mindful consumption, where one's purchases are intentional and necessary. A minimalist approach extends beyond personal benefit; it fosters a sense of satisfaction and fulfillment derived from experiences rather than accumulating items that often end up unused (Blackburn et al., 2023).

Practicing conscious consumerism further enhances sustainable living efforts. This means making informed purchasing decisions that support sustainable brands and discourage harmful production practices. By choosing products from companies that prioritize eco-friendly processes, consumers can drive market trends towards greater ecological responsibility. Conscious consumerism emphasizes the power of individual choices in shaping industry standards, fostering an environment where businesses are motivated to adopt greener practices due to consumer demand.

Taking these steps—reducing meat consumption, minimizing single-use plastics, adopting minimalism, and practicing conscious consumerism—demonstrates how minor adjustments in our daily lives can collectively lead to significant environmental benefits. These actions are not just about making sacrifices but about redefining what it means to live well in harmony with the planet. They empower individuals to take control of their impact on the environment, proving that everyone has a role to play in combating climate change.

Reducing meat consumption is particularly effective because the livestock industry is a major contributor to methane emissions. Methane is a greenhouse gas that is much more effective at trapping heat in the atmosphere than carbon dioxide, making it critical to address. Plant-based diets, which rely more on grains, vegetables, and fruits, have a smaller carbon footprint and require less energy and water to produce. Additionally, there are numerous health benefits associated with plant-based diets, including reduced risks of heart disease, hypertension, and certain cancers. Therefore, shifting dietary habits can simultaneously promote better personal health and environmental health.

The proliferation of single-use plastics poses a grave threat to marine life and ecosystems. Plastics take hundreds of years to decompose and often break down into microplastics, which contaminate waterways and enter the food chain. By opting for durable alternatives, such as metal straws, glass containers, and cloth grocery bags, individuals can drastically cut down the volume of plastic waste they generate. Not only does this help reduce landfill mass, but it also lessens the adverse effects on wildlife that often ingest or become entangled in plastic debris.

Minimalism offers an introspective path toward sustainable living by encouraging people to evaluate their consumption patterns. It challenges the contemporary narrative that equates happiness with possession accumulation. Minimalists focus on what truly brings value and joy into their lives, which often includes relationships, experiences, and personal growth rather than material goods. This shift not only reduces the environmental cost of producing and discarding unnecessary items but also aligns with psychological principles that suggest experiential purchases tend to bring longer-lasting happiness compared to material ones. The practice of minimalism can therefore foster a deeper sense of wellbeing while contributing to environmental preservation.

Conscious consumerism places a magnifying glass on supply chains and business operations, urging consumers to support brands that align with ethical and environmental values. This practice encourages transparency and accountability within industries, pushing companies to adopt greener and fairer practices. When consumers choose products from companies that prioritize sustainability, they indirectly support practices that reduce environmental harm and promote social good. This collective action sends a powerful message to industries, demonstrating that there is a market preference for products that do not come at the expense of the planet or marginalized communities.

Incorporating these lifestyle changes may seem daunting at first, but they become easier with gradual implementation and community support. Sharing tips, joining local sustainability groups, and encouraging friends and family to adopt similar practices can create a network of positive reinforcement. Over time, these small actions add up, leading to substantial reductions in carbon emissions and fostering a culture of environmental responsibility.

Energy-Efficient Practices in Households

Increasing energy efficiency in our homes is a practical and impactful way to contribute to environmental sustainability. By adopting specific steps, households can significantly reduce their energy consumption and emissions. These efforts not only benefit the planet but also lead to cost savings and enhanced comfort within the home.

Upgrade to Energy-Efficient Appliances

One of the most straightforward actions you can take is upgrading to energy-efficient appliances. Older models of refrigerators, dishwashers, washing machines, and other household devices often consume more energy than necessary. By replacing them with appliances that have high energy efficiency standards, such as those bearing the Energy Star label, you can lower your energy use and utility bills over time. For instance, an Energy Star-rated refrigerator uses up to 15% less energy than non-rated models. Not only do these appliances help reduce energy consumption, but they also result in long-term financial savings due to lower electricity costs. This approach demonstrates how a single change can make a significant difference.

Implement Smart Home Technology

Another effective measure is the implementation of smart home technology. Smart thermostats, for example, are designed to optimize heating and cooling schedules based on your daily routines. They learn your habits and adjust the temperature accordingly, ensuring that energy is not wasted when it is not needed. Moreover, these devices can be controlled remotely through smartphone apps, allowing you to manage your home's energy usage efficiently, even when you're not there. Home automation systems can further enhance this by integrating lighting, heating, and cooling controls, providing comprehensive energy management solutions. By leveraging such technologies, households can achieve substantial reductions in energy waste and associated costs.

Conduct Energy Audits

Understanding how your home consumes energy is crucial for identifying areas where improvements can be made. Conducting an energy audit involves evaluating various aspects of your home's energy use, including insulation quality, air leaks, and the efficiency of heating and cooling systems. Through professional or DIY audits, homeowners can uncover hidden issues that lead to energy loss.

For example, poor insulation or drafts around doors and windows can cause significant heat loss during winter and cooling inefficiencies in summer. Addressing these issues, such as by adding insulation or sealing gaps, can considerably improve your home's energy efficiency. An energy audit provides a clear roadmap for making targeted improvements that enhance overall performance and reduce emissions.

Utilize Renewable Energy Sources

Incorporating renewable energy sources into your home is a powerful step towards sustainable living. Solar panels, for instance, harness the sun's energy to generate electricity, decreasing reliance on the grid and reducing electricity costs. Installing solar panels can be a significant investment initially, but local incentives and long-term savings on energy bills make it a worthwhile consideration. Additionally, wind turbines can be another option for generating clean energy if the geographical location supports it. While these installations require an upfront investment, the benefits of generating your own clean energy extend far beyond cost savings. They represent a commitment to reducing carbon footprints and promoting environmental health.

The transition to renewable energy might seem daunting, but many communities offer programs and incentives to support homeowners in making this shift. Government rebates, tax credits, and other financial incentives can offset installation costs and make renewable energy solutions more accessible. The impact of utilizing renewable sources is profound, not just in reducing individual energy bills, but also in contributing to broader efforts to combat climate change.

Actionable Guidelines for Households

To implement these measures effectively, here are some guidelines for each actionable step:

1. **Upgrade to Energy-Efficient Appliances** :

   - Look for the Energy Star label when purchasing new appliances.

- Compare the energy consumption ratings of different models.

- Consider the long-term savings on energy bills when evaluating the cost.

1. **Implement Smart Home Technology** :

- Invest in a programmable or smart thermostat.

- Use automation systems that allow you to control lights, heating, and cooling remotely.

- Regularly update settings and schedules to match your household's routine.

1. **Conduct Energy Audits** :

- Start with a basic check of insulation and potential air leaks.

- Hire a professional auditor for a comprehensive assessment.

- Apply suggested improvements like adding insulation or sealing gaps promptly.

1. **Utilize Renewable Energy Sources** :

- Research local incentives and rebates for installing solar panels or wind turbines.

- Choose reliable and experienced contractors for installation.

- Monitor the performance of installed systems to ensure optimal operation.

Each of these steps represents a proactive approach to achieving greater energy efficiency in your home. By implementing these strategies, households can play a pivotal role in reducing energy consumption and carbon emissions. These efforts align closely with broader environmental goals and demonstrate a commitment to creating a sustainable future. Small changes in our daily lives, when combined, have the power to drive significant global impacts, highlighting the importance of every individual's contribution to protecting our environment.

Sustainable Transportation Options

In our journey towards sustainable living, one key aspect is transforming the way we commute. Sustainable transportation methods play a pivotal role in reducing our carbon footprint and promoting eco-friendly behaviors. Here's how you can make a difference:

Promote Public Transit Use

Public transportation systems like buses, trams, and subways are efficient and eco-friendly alternatives to personal vehicles. By relying on public transit, individuals can significantly reduce their carbon footprints. According to research, taking public transportation even for a few trips can diminish the carbon footprint of an average household by up to 6,000 pounds (3 tons) of CO2e annually (*Sustainable Transportation | Cool California*, n.d.). Opting for public transit not only helps the environment but also promotes efficient support for these vital community systems.

Using buses, trains, or subways reduces the number of vehicles on the road, decreasing traffic congestion and air pollution. It offers financial benefits as well, such as saving money on gas, parking fees, and vehicle maintenance. Moreover, public transit allows passengers to use their travel time productively, whether reading a book, catching up on work, or simply relaxing without the hassle of driving.

Encourage Cycling and Walking

Cycling and walking are exemplary forms of active transportation that bring immense benefits both to individuals and the environment. These modes of traveling are emission-free, making them the most sustainable choices for shorter distances. Communities that encourage cycling and walking see improved air quality and less reliance on motor vehicles, contributing to a safer and healthier environment for everyone.

In addition to environmental benefits, cycling and walking enhance physical health. Regular physical activity is associated with a lower risk of chronic diseases such as heart disease, diabetes, and obesity. Adults living in walkable neighborhoods tend to be more fit and have a reduced risk of diseases related to sedentary lifestyles. Encouraging safe infrastructure like bike lanes and pedestrian paths can further stimulate these healthy habits, fostering a sense of community engagement.

A practical example could be seen in urban areas where bike-sharing programs and pedestrian-friendly zones are implemented. Cities that prioritize bike lanes and secure pedestrian pathways not only foster healthier lifestyles but also attract tourism, boosting local economies.

Carpooling and Ridesharing

Carpooling and ridesharing are efficient ways to lessen the number of vehicles on the road, thereby reducing overall emissions and traffic congestion. Sharing rides with coworkers, friends, or through rideshare apps diminishes individual carbon footprints and enhances social connections. Carpooling can reduce the carbon footprint of an average household by up to 2,000 pounds (1 ton) of CO2e annually (*Sustainable Transportation | Cool California*, n.d.).

Participating in a carpool means fewer cars on the road, leading to less traffic and smoother commutes. Furthermore, carpool participants often gain access to High Occupancy Vehicle (HOV) lanes, which can save commuting time. For those dealing with long daily drives, this time-saving benefit can lead to reduced stress and increased productivity.

Companies and educational institutions can encourage carpooling by offering incentives such as preferred parking spots or organizing internal carpooling networks. Technological advancements have made it easier to find carpool partners through mobile applications, enhancing the feasibility and convenience of this option.

Explore Electric Vehicles (EVs)

Electric Vehicles (EVs) present a promising alternative to traditional gasoline and diesel-powered cars. EVs operate on electric power, which results in zero-emission driving, making them a highly sustainable choice. The reduction in greenhouse gas emissions from using EVs can be substantial. Households switching to electric cars can see significant reductions in their carbon footprint while taking advantage of government incentives and improved charging infrastructure.

The cost savings from operating an EV can also be compelling. While there might be a higher initial investment compared to conventional cars, the long-term savings on fuel and lower maintenance costs often outweigh the upfront expenses. Additionally, governments around the world offer various rebates and tax incentives to promote the adoption of electric vehicles, making them more affordable.

EV technology is rapidly advancing, resulting in longer ranges between charges and faster charging times. Public and private investments in charging stations are on the rise, enhancing the practicality of owning an EV. Resources such as MyGreenCar and UC Davis' Electric Vehicle Explorer provide valuable tools for prospective buyers to determine the best EV based on driving patterns and costs (*Sustainabl e Transportation | Cool California*, n.d.).

Promoting Community-Wide Sustainability

In a world increasingly aware of its environmental impact, encouraging community efforts to embrace sustainable practices is essential. When individuals within a community come together to adopt eco-friendly habits, the collective effort can result in significant positive changes for the environment. This subpoint delves into how educational resources, local business engagement, community projects, and success stories can drive broader community adoption of sustainable practices.

Educational Resources: One of the foundational steps towards fostering a sustainable community is providing access to comprehensive educational resources. Education allows individuals to understand the importance and impact of their actions on the environment. For instance, community centers can host workshops that demonstrate simple yet effective sustainable practices such as composting, recycling, and energy conservation. Libraries could curate a collection of books, pamphlets, and digital resources focused on environmental education.

Schools can integrate sustainability into their curriculum, teaching students about renewable energy, biodiversity, and waste management. By offering hands-on activities like building solar-powered devices or organizing nature walks, educators can make learning about sustainability engaging and practical. Public awareness campaigns using social media platforms can further disseminate knowledge, reaching a wider audience with tips and guidelines on living sustainably.

Local Business Engagement: Local businesses play a crucial role in promoting community-wide sustainable practices. Encouraging businesses to adopt eco-friendly measures not only reduces their carbon footprint but also sets a standard for the entire community. For example, a local grocery store might switch to biodegradable packaging and run campaigns encouraging customers to bring reusable bags. Such initiatives align with Goal 12 of the Sustainable Development Goals, which emphasizes responsible consumption and production (*PR on the GO Local Entrepreneurs and the UN's Sustainability Agenda – Strategies for Business Success*, 2024).

Businesses can also engage in local environmental projects. A hardware store could partner with municipal authorities to beautify public spaces by planting trees and flowers, contributing to urban greening efforts. Restaurants might source ingredients locally to reduce transportation emissions and support regional farmers, thereby promoting a more sustainable food system. By participating in or sponsoring community clean-up events, local enterprises not only help maintain a cleaner environment but also build a sense of communal responsibility.

Community Projects: Collective community projects are an excellent way to promote cooperation and shared responsibility among residents. Community gardens are a prime example, transforming vacant lots into productive green spaces where residents can grow their own fruits and vegetables. These gardens provide fresh, local produce while educating participants about sustainable agriculture and reducing the need for long-distance transportation of food items.

Another impactful project is organizing regular community clean-up drives. These events encourage residents to take an active role in maintaining their local environment, picking up litter, and properly disposing of waste. Clean-up initiatives can be expanded to include recycling programs where residents can learn to sort recyclable materials correctly, as well as electronic waste collections to safely dispose of gadgets and batteries.

Workshops can be held to teach residents about rainwater harvesting, energy-efficient home improvements, and other ways to live sustainably. Creating a community composting program can also divert organic waste from landfills while producing nutrient-rich soil for gardens. Involving local schools, businesses, and organizations in these projects ensures broader participation and reinforces the message that everyone has a role to play in sustainability.

Success Stories: Sharing examples of successful community sustainability initiatives can be incredibly motivating. Highlighting case studies where communities have made significant strides toward sustainability demonstrates that positive change is possible and encourages others to follow suit. For instance, a neighborhood that successfully transitioned to zero-waste living by implementing thorough recycling programs, composting, and reducing single-use plastics can serve as an inspiring model.

Communities can celebrate their achievements through various channels, such as newsletters, social media posts, and local news features. Hosting events where community members share their experiences and strategies for sustainable living fosters a spirit of collaboration and learning. Awards and recognition for outstanding contributions to sustainability can also incentivize individuals and groups to innovate and take action.

For example, a coastal town that reduced ocean pollution by banning plastic bags and straws, while promoting beach clean-ups and installing proper waste disposal units, can inspire similar actions in other waterfront communities (*PR on the GO Local Entrepreneurs and the UN's Sustainability Agenda – Strategies for Business Success*, 2024). Another community might create a thriving urban forest through concerted tree-planting efforts, demonstrating how collective action can significantly enhance local biodiversity and environmental health.

Promoting these success stories through engaging storytelling techniques makes them relatable and accessible. Documentaries, blog posts, podcasts, and interactive community meetings can all be used to share these narratives, making it easy for others to understand the steps taken and the benefits achieved. This kind of sharing not only provides practical roadmaps but also strengthens the sense of community around a common goal of sustainability.

Final Thoughts

This chapter has explored practical ways we can change our daily habits to live more sustainably. By reducing meat consumption, minimizing the use of single-use plastics, adopting minimalism, and practicing conscious consumerism, we can each play a part in protecting our planet. These small adjustments not only lower our carbon footprints but also help us lead healthier and more fulfilling lives. The combined effort of these actions shows that everyone has a role in fighting climate change.

It's important to remember that these lifestyle changes don't have to be made all at once. Gradually incorporating these practices into your routine can make the transition smoother and more manageable. Sharing these tips with friends and family or joining local sustainability groups can create a supportive community that motivates you to stick with these habits. Collectively, our individual efforts can lead to significant environmental benefits, proving that even small steps can make a big difference in the fight against climate change.

Chapter 6

Community Resilience: Building Local Solutions

Building local solutions to bolster community resilience is a vital step in facing the challenges posed by climate change. Community resilience focuses on strengthening the capacity of local areas to not only adapt to but also thrive amidst environmental changes. Central to this endeavor are initiatives that harness collective effort, foster self-reliance, and promote sustainable practices at the grassroots level. The importance of these collaborative efforts cannot be overstated; they serve as the bedrock for creating robust systems capable of withstanding and recovering from climate-induced disruptions. The chapter delves into various strategies and examples that illustrate how communities can come together to craft effective and sustainable responses to environmental challenges.

In this chapter, we will explore numerous avenues through which communities can enhance their resilience. One significant aspect covered includes the development of local food production systems, such as community gardens and farmers' markets, which crucially reduce dependence on external food networks. Additionally, the chapter examines the role of cooperative models in promoting sustainable farming practices and resource sharing, which collectively contribute to long-term agricultural viability. Furthermore, the chapter sheds light on the educational programs that empower individuals with knowledge about sustainable agriculture and food preservation techniques. These programs not only enrich personal lives but also bolster community food security. By immersing in these topics, readers will gain a deeper understanding of how communal efforts and local solutions can pave the way for resilient futures.

Local Food Production and Security

Local food production plays a critical role in enhancing food security and building resilience within communities, particularly those facing the impacts of climate change. By sourcing food locally, communities can reduce their dependence on external food systems and create sustainable, self-reliant networks. One effective approach to local food production is through community gardens, which empower residents and foster shared responsibilities.

Community gardens serve as educational hubs where residents can learn about sustainable agriculture practices. These gardens offer hands-on experiences in growing food, allowing novices to gain confidence and skills under the guidance of more experienced gardeners. This shared responsibility strengthens community bonds and promotes a sense of ownership, encouraging collective action towards common goals. As residents work together to cultivate these spaces, they also engage in a form of social learning that enhances trust and cooperation among community members (C. et al., 2020).

In addition to their educational value, community gardens address food insecurity by providing access to fresh produce. For many urban areas with limited access to supermarkets, these gardens can significantly improve the availability of healthy food options. Esther, a Food Equity Advisor, highlighted the potential for community-run gardens to overcome barriers of affordability and distance, making fresh fruits and vegetables accessible to all residents (Colson-Fearon & Versey, 2022).

Farmers' markets are another vital component of local food production, promoting economic resilience and reducing environmental impact. By enabling direct sales between farmers and consumers, these markets strengthen local economies. Farmers benefit from fairer prices due to the elimination of middlemen, while consumers gain access to high-quality, fresh produce. Moreover, farmers' markets encourage seasonal eating, which lowers the carbon footprint associated with long-distance transportation and storage of food.

Beyond economic benefits, farmers' markets serve as venues for education and community engagement. Shoppers at these markets often prioritize access to fresh produce but also value the social interactions and the opportunity to learn more about where their food comes from (Lockeretz, 1986). By fostering connections between farmers and consumers, these markets promote a greater understanding of agricultural practices and the importance of supporting local food systems.

Cooperative models further enhance food security and resilience by pooling resources and sharing innovative agricultural practices. Cooperatives provide a platform for farmers to collaborate, ensuring fair prices for their produce while keeping food affordable for consumers. These models enable farmers to leverage collective knowledge and resources, facilitating the adoption of new techniques to counteract climate challenges. For example, cooperatives can invest in sustainable farming practices such as crop rotation, organic fertilization, and water-efficient irrigation systems, which contribute to long-term agricultural viability.

Furthermore, cooperatives often include provisions for resource sharing, such as equipment and storage facilities, which can reduce individual costs and increase overall efficiency. By working together, farmers can mitigate risks and adapt more effectively to changing environmental conditions. This collaborative approach not only supports local agriculture but also strengthens community ties, fostering a culture of mutual support and resilience.

Educational programs play a crucial role in raising awareness about sustainable farming, preservation techniques, and the benefits of local sourcing. These programs empower consumers to make informed choices that support local food systems and sustainability. For instance, workshops on home gardening, composting, and food preservation teach valuable skills that enhance personal and community food security. Educational initiatives can also include farm tours, cooking classes, and informational sessions at farmers' markets, offering diverse ways for individuals to engage with and learn about their local food systems.

Additionally, educational programs can highlight the broader societal and environmental impacts of food choices. By understanding how food production affects ecosystems, resource use, and climate change, consumers are better equipped to make decisions that align with their values and contribute positively to their communities. These programs can also foster advocacy and civic engagement, encouraging individuals to participate in policy discussions and initiatives that support sustainable agriculture and food security (Smith et al., 2019).

Community-Based Renewable Energy Projects

Local renewable energy initiatives offer powerful solutions for sustainable energy use and community empowerment. One compelling example of this is solar cooperatives. These cooperatives pool community resources to cover the installation costs of solar panels, making renewable energy more accessible. By doing so, they not only democratize access to clean energy but also drive economic growth by creating jobs in the renewable energy sector. For instance, members of a solar cooperative share both the benefits of reduced energy costs and the responsibilities of maintenance, which fosters a sense of collective ownership and commitment to sustainability.

Solar cooperatives also serve as educational hubs, where community members can learn about solar technology and its benefits. By engaging in these projects, residents gain valuable skills that increase their employability in the growing green jobs market. Additionally, solar cooperatives often collaborate with local schools and colleges, offering internships and training programs that help students prepare for careers in renewable energy. This creates a cycle of education and employment that strengthens the entire community.

Moreover, small-scale community wind projects offer another avenue for sustainable energy production tailored to local needs. These projects harness wind energy to generate electricity, reducing reliance on traditional fossil fuels and cutting down greenhouse gas emissions. In areas with consistent wind patterns, community-owned wind turbines can provide a reliable source of power, making communities less vulnerable to energy price fluctuations. Wind projects also present opportunities for local job creation during the construction and maintenance phases, further contributing to economic resilience.

For example, rural areas that invest in wind energy can benefit from the stabilization of energy costs over time. As the initial investment in wind turbines is recouped, the cost of energy production decreases, providing long-term financial benefits to the community. This stability is particularly important in regions where energy prices are subject to market volatility, helping to shield residents from sudden increases in energy costs.

Community microgrids represent another innovative approach to enhancing local energy resilience. Microgrids are localized networks that operate independently from the main grid, using a combination of renewable energy sources like solar and wind, supplemented by energy storage systems such as batteries. By optimizing energy usage and reducing transmission losses, microgrids make energy distribution more efficient and reliable.

The development of microgrids offers multiple benefits to a community. Firstly, they increase energy security by providing a backup power source during outages. This is especially critical in areas prone to natural disasters, where maintaining power can be a matter of life and death. Secondly, microgrids can be tailored to meet the specific energy needs of a community, ensuring that resources are used effectively and sustainably.

Furthermore, community involvement in the development and management of microgrids boosts local investment and pride. Residents who participate in these projects often feel a stronger connection to their community and a greater sense of responsibility towards its well-being. The economic benefits are also significant; as the community invests in its own energy infrastructure, the profits and savings are reinvested locally, fostering economic growth and resilience.

Energy efficiency programs complement renewable energy initiatives by educating residents on ways to reduce consumption and costs. These programs advocate for practical home upgrades, such as installing energy-efficient appliances, improving insulation, and adopting smart home technologies. By making homes more efficient, residents can significantly lower their energy bills while reducing their overall carbon footprint.

Educational campaigns are a vital component of these programs. Workshops, seminars, and online resources provide residents with the knowledge and tools they need to implement energy-saving measures. Engaging community leaders and local organizations in these efforts helps to spread awareness and encourage widespread participation. When energy efficiency becomes a community-wide effort, it fosters a culture of sustainability that benefits everyone.

One practical guideline for establishing successful solar cooperatives is to begin with a thorough assessment of community needs and resources. Engaging stakeholders early in the process ensures that the initiative addresses local priorities and gains broader support. Additionally, securing grants and other funding sources can help offset initial costs, making the cooperative more financially viable.

For small-scale wind projects, it's essential to conduct rigorous site assessments to determine the feasibility of wind energy in the area. Community consultations and transparent planning processes can help address any concerns and build local support. Partnering with experienced developers and engineers can also ensure that the project is technically sound and economically sustainable.

When developing community microgrids, it's crucial to involve a diverse group of stakeholders, including local government, businesses, and residents. Collaborative planning and decision-making can lead to more equitable and effective outcomes. Additionally, securing technical expertise and reliable funding sources can help navigate the complexities of microgrid implementation.

Lastly, energy efficiency programs should prioritize accessibility and inclusivity. Offering incentives for low-income households and providing easy-to-understand information can help ensure that all community members can participate. Regularly updating the program based on feedback and new technologies can also keep it relevant and effective.

Disaster Preparedness and Response Plans

In the face of increasing climate change impacts, developing robust disaster preparedness and response frameworks within communities is essential. This section outlines key strategies to equip communities with the tools they need to face disasters effectively and recover swiftly.

Risk assessments are foundational in crafting tailored preparedness strategies. By evaluating local vulnerabilities—be it flood-prone regions, wildfire zones, or areas susceptible to storms—communities can design specific plans that address these unique threats. Engaging community members in the risk assessment process not only provides a diverse perspective on potential risks but also fosters a sense of collective responsibility. When people actively contribute to identifying their community's vulnerabilities, they are more likely to participate in and support subsequent preparedness measures. Additionally, these assessments highlight areas where resources may be lacking, allowing for strategic allocation in advance.

Emergency response training plays a crucial role in preparing residents to act efficiently during crises. Such training programs should cover a broad spectrum of skills, including first aid, search and rescue operations, and fire fighting techniques. Equipping residents with these skills ensures that they can assist each other before professional responders arrive, thus reducing immediate harm and chaos. These programs also help build a network of trained responders who are familiar with community-specific needs and resources. For instance, knowing the locations of shelters, medical facilities, and resource distribution centers can expedite emergency efforts. Regularly conducting these trainings helps refresh skills and introduces new community members to essential response protocols.

Effective preparedness plans must encompass clear communication channels, well-defined roles, and designated resource allocations. Communication is vital during emergencies; having a pre-established system ensures information flows quickly and accurately, preventing misinformation and panic. Resource allocations should outline the availability and distribution of essentials such as food, water, medical supplies, and shelter. Assigning roles ensures everyone knows their responsibilities, which streamlines efforts and avoids duplication of tasks. Regular drills are an integral part of maintaining these plans. Through drills, communities can test the practicality and relevance of their strategies, making necessary adjustments based on observations and feedback. Drills also reinforce familiarity with procedures, ensuring that actions become almost instinctive during real emergencies.

Collaboration with emergency services and NGOs enhances the overall preparedness and response capabilities of a community. These partnerships bring additional expertise, resources, and support, which are invaluable during disasters. Emergency services, including police, fire departments, and medical teams, offer critical intervention skills and equipment. NGOs often provide supplementary resources and specialized knowledge, such as psychological support and community rebuilding initiatives. Building relationships with these entities before disasters strike establishes trust and ensures smoother coordination when needed most. Joint exercises and planning sessions foster mutual understanding and enable better integration of efforts during actual emergencies.

Incorporating these strategies into a cohesive framework strengthens community resilience, enabling quicker recovery and reducing long-term impacts. Communities that engage in comprehensive risk assessments can develop targeted responses that mitigate identified threats. Emergency response training builds a capable population ready to assist immediately after disasters, bridging the gap until professional help arrives. Effective preparedness plans that include regular drills keep emergency protocols up-to-date and actionable. Collaborations with emergency services and NGOs expand a community's capacity to handle crises by pooling resources and expertise.

By embedding these elements into community preparedness frameworks, towns and cities can better withstand the pressures of climate-induced disasters. The proactive involvement of residents in assessing risks and participating in response training creates a knowledgeable and empowered populace. Clear communication, resource management, and role assignments streamline emergency operations, while regular drills ensure readiness. Partnerships with external agencies enhance support structures, providing additional layers of security and assistance.

In practical terms, communities can begin by conducting thorough risk assessments. Local governments can facilitate workshops and meetings to gather input from residents about perceived risks and historical data on past events. Tools such as hazard maps and vulnerability indices can guide these discussions, offering visual aids to pinpoint high-risk areas. Following the assessment, emergency response training programs can be organized, inviting experts to teach essential skills. Community centers, schools, and local organizations can serve as venues, encouraging broad participation.

Preparedness plans need to be drafted with input from various stakeholders, including local leaders, emergency services, and community groups. These plans should be documented and widely distributed, ensuring all residents have access to critical information. Establishing a dedicated communication platform, such as a community alert system via text messages or social media, ensures rapid dissemination of updates during emergencies. Regular drills can be scheduled quarterly or biannually, incorporating different scenarios to test various aspects of the response plan.

To forge strong collaborations, communities can reach out to local emergency services and NGOs to establish formal partnerships. Memorandums of Understanding (MOUs) can outline the scope of cooperation, resource sharing, and joint training initiatives. Inviting representatives from these organizations to community meetings and drills builds rapport and improves coordination.

Local Empowerment Through Sustainable Practices

Sustainable practices are powerful tools that can significantly enhance community resilience. By integrating these methods, communities can build robust local solutions to face climate change and its accompanying challenges. One of the most promising sustainable practices is the establishment of community gardens. These gardens do more than just produce food; they foster social connections among neighbors and create a sense of ownership and pride within the community. When residents come together to cultivate a shared space, they not only reap fresh produce but also develop strong bonds and a collective identity. This unity is crucial when facing environmental challenges as it ensures that community members support one another in times of need.

Community gardens serve as vibrant spaces for interaction and learning. Residents from diverse backgrounds can share gardening techniques and knowledge, creating a melting pot of agricultural wisdom. For instance, older generations often pass down traditional methods, while younger participants might introduce innovative ideas such as vertical gardening or hydroponics. These interactions enrich the community's collective expertise and ensure that gardening practices remain relevant and effective in changing climates.

Moreover, as people work together, they engage in physical activity, which promotes health and well-being, reducing the overall stress levels within the community.

Local farmers' markets play an equally vital role in promoting sustainable practices and community engagement. These markets provide an excellent venue for education about sustainable farming and encourage community members to buy locally produced food. Purchasing food from local farmers reduces transportation emissions and supports the local economy. At these markets, farmers often share insights into their sustainable practices, such as crop rotation, organic farming, and water conservation techniques. Shoppers learn firsthand about the benefits of supporting local agriculture and are more likely to make environmentally conscious decisions in their daily lives. Additionally, farmers' markets bring people together, creating a lively atmosphere where residents can connect, exchange ideas, and form lasting relationships.

Beyond selling produce, farmers' markets can host workshops and demonstrations on topics like composting, rainwater harvesting, and permaculture. These educational activities equip community members with practical skills to implement at home. They also highlight the importance of sustainable living, encouraging individuals to adopt eco-friendly habits that collectively make a significant impact. Through these efforts, farmers' markets become hubs of sustainability education and community involvement, reinforcing the local food system's resilience against climate-induced disruptions.

Cooperative farming initiatives offer another avenue for enhancing community resilience through sustainable practices. By pooling resources and sharing knowledge, cooperative farms can innovate and adapt to climate-induced agricultural challenges more effectively than individual farmers working in isolation. These cooperatives allow small-scale farmers to access expensive equipment, bulk-buy seeds and fertilizer, and invest in advanced technologies that would be unaffordable independently. This collaborative approach enables farmers to experiment with new methods, such as drought-resistant crops or integrated pest management systems, improving overall productivity and sustainability.

Furthermore, cooperative farming fosters a sense of solidarity and mutual support. In times of crisis, such as extreme weather events, cooperative members can rely on each other for assistance, whether it be sharing labor, equipment, or strategies for recovery. This network of support strengthens the community's ability to withstand and bounce back from adverse conditions, ensuring long-term resilience. Cooperative farming also provides a platform for continuous learning and adaptation, as farmers regularly exchange experiences and insights, staying abreast of best practices and emerging challenges.

Educational programs focused on food resilience are essential for empowering consumers to make informed choices that support the local food system and sustainability. These programs educate individuals about the origins of their food, the processes involved in its production, and the importance of sustainable practices. By understanding the journey from farm to table, consumers gain a deeper appreciation for locally produced food and its impact on their health and the environment. Programs may include cooking classes that emphasize seasonal ingredients, workshops on preserving and storing food, and seminars on the nutritional benefits of various crops.

Empowered with this knowledge, consumers are more likely to support local farmers and sustainable agriculture practices. They become advocates for the local food system, spreading awareness and encouraging others to make environmentally conscious choices. Educational programs also highlight the interconnectedness of food systems, demonstrating how individual actions contribute to broader environmental outcomes. This holistic understanding fosters a culture of sustainability, where community members actively participate in creating resilient food systems that can adapt to climate change.

The U.S. Department of Agriculture's People's Garden Initiative exemplifies how community efforts in gardening can align with broader sustainability goals. Launched to include school gardens, urban farms, and small-scale agriculture projects, this initiative emphasizes collaboration, conservation practices, and public education. Gardens that meet these criteria receive recognition and support, fostering a nationwide movement towards resilient local food systems (USDA Opens People's Garden Initiative to Gardens Nationwide, 2022).

Likewise, organizations like Fresh Approach work tirelessly to connect farmers and communities through nourishing food, promoting healthy food access, and engaging in urban agriculture. Their mission to create long-term change in local food systems by building resilient food and farming networks illustrates the transformative power of community-led initiatives. By involving diverse stakeholders, including nonprofits, government agencies, and health systems, Fresh Approach enhances food security and sustainability across California (Who We Are – Fresh Approach, n.d.).

Summary and Reflections

As we reflect on the strategies discussed, it becomes clear that enhancing community resilience through local efforts is both practical and powerful. From community gardens to farmers' markets and cooperative farming, these initiatives not only bolster food security but also nurture a sense of belonging and collective action. By working together to cultivate local resources, communities can reduce dependence on external food systems and foster sustainable practices that are essential in mitigating the impacts of climate change.

Moreover, education plays a pivotal role in this journey. Whether through workshops at farmers' markets or hands-on learning in community gardens, these educational opportunities empower individuals with the knowledge and skills needed to support local food systems and sustainability. As residents become more informed about sustainable agriculture and energy practices, they contribute to building a resilient community capable of facing environmental challenges head-on. The combined efforts of growing food locally, supporting sustainable markets, and sharing knowledge create a robust framework for community resilience, ensuring a healthier and more sustainable future for all.

Chapter 7

Navigating Policy and Advocacy: Driving Systemic Change

Navigating policy and advocacy is like steering a ship through complex waters. The way countries respond to climate change isn't just shaped by individual efforts but by the policies they enact and the activism driving those policies. Here, we explore how these elements intersect to create significant, systemic changes. Policies at an international level, such as treaties and agreements, provide a unified direction for nations to follow. Alongside this structured governmental action, grassroots movements bring a dynamic force that can influence public opinion and hold leaders accountable. Both avenues are essential for addressing a challenge as vast and interwoven as climate change.

This chapter delves into various facets of policy and advocacy in the context of climate change. We'll start by examining pivotal climate agreements like the Kyoto Protocol and the

Paris Agreement, which have set the stage for global cooperative efforts. Further on, we will discuss the importance of international collaboration in achieving climate targets and how national policies are influenced by these global commitments. You'll also read about the critical role that non-governmental organizations play in advocating for stronger environmental laws and representing marginalized communities. Finally, we'll look at the challenges faced by countries in implementing these policies and explore future directions for effective climate action.

Climate policies and international agreements

Global policies and treaties play an essential role in creating frameworks that direct national and local climate actions. They lay the groundwork for synchronized efforts, facilitating a comprehensive approach to addressing climate change on multiple levels.

Historical Overview of Pivotal Climate Agreements

One of the earliest and most significant global climate agreements is the Kyoto Protocol, adopted in 1997 and coming into force in 2005. It was the first legally binding climate treaty, compelling developed countries to reduce their greenhouse gas emissions by an average of 5 percent below 1990 levels. Despite its groundbreaking nature, the Kyoto Protocol faced criticism for not obligating developing countries, including major emitters like China and India, to take action. The United States' decision to withdraw further weakened the protocol's impact (Maizland, 2023). Despite these challenges, the Kyoto Protocol set a vital precedent for future international climate commitments.

A more universally encompassing agreement is the Paris Agreement, adopted in 2015. Unlike the Kyoto Protocol, the Paris Agreement requires all countries to set their own emissions-reduction targets known as nationally determined contributions (NDCs).

The primary goal is to keep the global temperature increase well below 2°C above pre-industrial levels, with aspirations to limit the rise to 1.5°C. This agreement marks a significant shift towards global net-zero emissions in the latter half of this century. The collaborative nature of this agreement has made it a cornerstone in the fight against climate change and a critical framework for guiding national policies (United Nations, 2015).

Importance of International Cooperation

International cooperation is paramount in setting ambitious climate targets. Climate change is a global issue that transcends borders, necessitating collective action. The success of the Paris Agreement hinges on the willingness of countries to cooperate and commit to progressively bolder climate actions every five years. By working together, nations can share resources, technology, and strategies to mitigate and adapt to climate change effectively.

Moreover, international summits like the annual Conference of the Parties (COP) provide a platform for countries to assess progress, share best practices, and strengthen their commitments. The iterative review process embedded within the Paris Agreement ensures a mechanism that encourages transparency and accountability. This cooperative effort also extends to financial mechanisms aimed at supporting developing countries in their climate actions, acknowledging that while climate change affects all, the capacity to respond varies significantly across nations (United Nations, 2015).

Challenges Faced by Countries

Despite the clear benefits of international climate agreements, countries encounter numerous challenges in committing to and implementing these agreements. One significant hurdle is the political and economic landscape within individual nations. For instance, shifting from fossil fuels to renewable energy sources requires substantial investments and policy shifts that may face opposition from powerful industrial lobbies.

Furthermore, there are disparities in how climate change impacts different regions, leading to conflicting priorities. Developing countries often grapple with immediate socio-economic issues, making long-term climate strategies a lower priority. Additionally, the issue of fairness arises, particularly around historical emissions and the principle of common but differentiated responsibilities. Developed nations have historically contributed more to greenhouse gas emissions, while developing nations argue for their right to industrialize and improve living standards without bearing undue climate mitigation burdens (Maizland, 2023).

Impact on National Policies

Global climate agreements profoundly influence national policies. They act as catalysts for governments to integrate climate goals into national legislation. For example, after ratifying the Paris Agreement, many countries, including the European Union member states, formulated ambitious climate plans aiming for net-zero emissions by mid-century. These plans often encompass regulations on energy production, transportation, industry emissions, and even agriculture.

Additionally, adherence to international treaties prompts the establishment of monitoring and reporting systems to track progress toward national climate targets. This systematic approach not only enhances transparency but also enables countries to adjust policies based on performance and emerging scientific insights. The ripple effect of global agreements often leads sub-national entities, like state or municipal governments, to develop localized climate action plans aligned with broader national and international goals.

For instance, in response to the Paris Agreement, many U.S. states and cities pledged to uphold the accord's principles even after the federal government announced its withdrawal. These local commitments underscore the cascading influence of international treaties on driving climate action across governance levels.

Future Directions for Climate Policy

Looking ahead, it is crucial to continue fostering international collaboration while ensuring that climate policies evolve to meet escalating challenges. One promising direction involves bolstering the resilience of communities by integrating climate adaptation measures into development planning. This includes investing in sustainable infrastructure, enhancing disaster preparedness, and protecting natural ecosystems.

Another critical area is the acceleration of the transition to renewable energy. Governments must implement policies that incentivize the adoption of clean energy technologies, such as solar, wind, and hydroelectric power. This transition not only mitigates greenhouse gas emissions but also creates economic opportunities through new industries and job creation.

Effective climate policy must also address social equity concerns. Ensuring that vulnerable populations, who are often the most impacted by climate change, receive adequate support and resources is imperative. Policymakers should strive to create inclusive strategies that promote environmental justice and equitable access to the benefits of climate action.

Lastly, continuous innovation and technological advancement will play a pivotal role in shaping future climate policies. Investing in research and development can unlock new solutions for carbon capture, storage, and utilization, reducing emissions from hard-to-abate sectors like heavy industry and aviation.

Role of governments and non-governmental organizations

In addressing climate change, the collaboration between government bodies and non-governmental organizations (NGOs) is paramount. These partnerships are essential for implementing effective climate solutions and policies. This section delves into the intricate dynamics of these collaborative efforts, illustrating the roles and impacts of governmental regulations, NGO advocacy, their successful partnerships, and the importance of monitoring outcomes through measurable goals.

Governments play a crucial role in the regulation and enforcement of environmental policies. They are responsible for creating frameworks that set the standards for environmental protection and sustainable practices. Governments have the authority to enact laws and regulations that limit carbon emissions, promote renewable energy, and ensure compliance with international climate agreements. Their role extends to funding and supporting research and development of new technologies and initiatives aimed at mitigating climate change. By imposing penalties for non-compliance and offering incentives for green practices, governments can drive significant change across various sectors.

However, governments alone cannot tackle the vast challenges posed by climate change. NGOs act as catalysts for change by influencing policy and public opinion. These organizations bring expertise, passion, and grassroots mobilization to the table. NGOs often fill gaps left by governmental policies and engage in activities such as raising awareness about climate issues, educating the public, and advocating for stronger environmental laws. For instance, campaigns led by NGOs have successfully pressured governments to adopt more stringent emissions targets and invest in sustainable infrastructure. NGOs also play a critical role in representing marginalized communities whose voices might otherwise be overlooked in policy discussions.

Partnerships between governments and NGOs have proven particularly effective in driving systemic change. One notable example is the collaboration on child-centred disaster risk reduction and climate change adaptation in Indonesia from 2008 to 2019. This long-term observation highlighted how school safety programs' sustainability depended on both government commitments and the nudging strategies employed by NGOs (Lassa et al., 2023). By combining efforts, they managed to institutionalize these programs at various governance levels, ensuring a broader and more lasting impact.

Another example can be seen in how environmental NGOs have partnered with local governments to implement community-based renewable energy projects. These initiatives not only reduce carbon footprints but also increase energy access in underserved areas. Through joint efforts, NGOs have been able to leverage government resources and influence policy changes that support the scaling up of successful local projects. Such collaborations demonstrate the power of uniting different strengths and resources to achieve common environmental goals.

To ensure the success of these collaborative efforts, it is crucial to track their effectiveness using measurable goals and indicators. This involves setting clear targets for environmental outcomes and regularly assessing progress against these benchmarks. Tracking mechanisms might include measuring reductions in greenhouse gas emissions, increases in renewable energy usage, or improvements in biodiversity conservation. By utilizing data and evidence-based approaches, both governments and NGOs can adjust their strategies as needed to enhance effectiveness.

For example, tracking the impact of reforestation projects could involve monitoring tree survival rates, carbon sequestration levels, and the socio-economic benefits to local communities. Transparency in these measurements allows stakeholders to understand what works and where improvements are necessary. Additionally, it fosters accountability, ensuring that both governmental bodies and NGOs remain committed to their objectives.

Moreover, setting realistic and achievable goals helps maintain momentum and encourages continuous improvement. It provides a framework for learning and adaptation, which is essential in the ever-evolving field of climate action. Effective tracking mechanisms also facilitate better communication with the public, helping to build trust and support for ongoing and future initiatives.

Guidelines for setting measurable goals should focus on SMART criteria—specific, measurable, achievable, relevant, and time-bound. This ensures that each goal is clearly defined and progress can be objectively evaluated. Regular reviews and reports on these metrics will help identify successes and challenges, enabling partners to refine their approaches continuously.

Gaps and challenges in climate policy

Exploring areas where policy has lagged behind scientific recommendations is crucial in understanding the challenges faced in addressing climate change effectively. One major issue is the disconnect between policy goals and the real-world implementation of climate strategies. Often, policies are designed with ambitious targets that align with scientific recommendations but fail to translate into actionable plans. For example, while many countries have committed to reducing carbon emissions, the actual measures taken often fall short of these commitments due to lack of proper infrastructure, funding, and political will.

Socio-political barriers also play a significant role in impeding effective climate action. Political interests and economic priorities often overshadow environmental concerns, leading to policies that prioritize short-term economic gains over long-term sustainability. For instance, powerful lobbying groups representing fossil fuel industries can influence policymakers to adopt less stringent regulations, thereby delaying the transition to renewable energy sources. Additionally, political instability in some regions can result in inconsistent climate policies, as successive governments may have differing priorities and approaches to climate action.

The need for more equitable policies that address social justice alongside environmental concerns cannot be overstated. Climate change disproportionately affects marginalized communities, including low-income households, indigenous populations, and people of color. Equitable policies should ensure that these communities are not only protected from the adverse effects of climate change but also benefit from climate solutions. For example, transitioning to renewable energy should include job creation programs that prioritize workers from communities affected by the decline of the fossil fuel industry. Moreover, policies should be designed to avoid unintended consequences, such as higher public transport fares that disproportionately impact poorer households (World Bank, 2023).

One example illustrating the critical gaps in current climate policies is the Paris Agreement. While it was a landmark achievement in setting global climate goals, its implementation has been uneven across countries. Many signatories have struggled to meet their nationally determined contributions (NDCs) due to a lack of clear roadmaps, insufficient funding, and inadequate enforcement mechanisms. Similarly, policies aimed at reducing deforestation through initiatives like REDD+ (Reducing Emissions from Deforestation and Forest Degradation) have faced challenges. Indigenous communities, who depend on forests for their livelihoods, are often excluded from decision-making processes, resulting in policies that do not adequately address their needs and could potentially harm their way of life (World Bank, 2023).

Guidelines for building coalitions and networks can help bridge the gap between policy goals and their implementation. Collaborative efforts between governments, NGOs, and grassroots organizations can enhance the effectiveness of climate policies. For instance, creating partnerships with local communities ensures that policies are culturally appropriate and address the specific needs of those most affected by climate change. Moreover, fostering collaboration among various stakeholders can lead to more comprehensive and inclusive climate strategies that leverage diverse forms of knowledge, including scientific, indigenous, and local expertise (Cairney et al., 2023).

Harnessing technology and social media is another powerful tool for overcoming socio-political barriers. Digital platforms can raise awareness about climate issues, mobilize public support, and hold policymakers accountable.

Innovative technological solutions can also enhance the monitoring and reporting of climate actions, ensuring transparency and accountability. For example, satellite technology can track deforestation in real-time, enabling immediate interventions to prevent illegal logging. Social media campaigns can amplify the voices of marginalized communities, highlighting their experiences and advocating for more inclusive climate policies.

Advocacy strategies for citizen involvement

In today's world, the pressing nature of climate change demands immediate and sustained action from individuals at all levels of society. Advocacy strategies empower citizens to influence policies that address these environmental challenges effectively. Engaging in advocacy can seem daunting, but understanding how individual actions can coalesce into significant movements is crucial.

Individual actions are often the catalyst for larger movements, especially when efforts are localized. A single person's commitment can inspire a community, which then has the potential to grow into a broader movement. Imagine participating in a local clean-up drive; this simple act not only improves the local environment but also raises awareness among neighbors. Sharing your experiences through neighborhood newsletters or local social media groups can amplify the impact, making others more likely to join the cause, thereby creating a snowball effect of positive change.

Localized efforts don't stop at clean-ups. Planting trees, participating in community gardens, or advocating for local policy changes can contribute to mitigating climate change. For example, urging your city council to adopt greener waste management practices or install solar panels on public buildings can have lasting effects. These localized efforts often resonate more powerfully because they are tailored to the specific needs and capabilities of the community.

Collective action is another pillar of effective advocacy, enhancing the power of individual efforts exponentially. When people come together, they amplify their voices, making it harder for policymakers to ignore their demands. Grassroots movements exemplify this collective power. One notable example is the Fridays for Future campaign started by Greta Thunberg. Her solo school strike evolved into a global movement, with millions of students worldwide participating and demanding action on climate change (University of Minnesota, 2023).

The significance of collective action is also evident in historical successes. For instance, the civil rights movement in the United States showcased how united efforts could lead to substantial legislative changes. Similar tactics are used in environmental advocacy, where petitions, marches, and demonstrations draw attention to pressing issues and pressure governments to act. Collective actions, therefore, create a sense of urgency and community, which are both essential for driving systemic change.

Modern tools for advocacy, particularly social media and online campaigns, are pivotal in contemporary climate activism. Social media platforms like Facebook, Instagram, Twitter/X, TikTok, and YouTube offer unprecedented opportunities to reach wide audiences quickly and effectively. These tools enable activists to share information, organize events, and mobilize supporters effortlessly. A well-crafted social media post can go viral, drawing attention from around the globe and encouraging diverse groups to rally behind a cause.

Creating engaging content is vital. Visual aids such as infographics, photos, and videos are powerful tools to convey messages succinctly and compellingly. Videos, in particular, can be extremely persuasive. Consider creating short documentaries or personal stories about how climate change affects lives. Using social media's interactive features, like polls and live streams, allows for real-time engagement with your audience, fostering a deeper connection and understanding.

Successful citizen advocacy movements provide inspiring examples of how grassroots efforts can lead to policy changes. The Dakota Access Pipeline protests showed how indigenous communities and environmental activists could unite to protect land and water resources. Despite facing numerous challenges, their persistence led to temporary halts in pipeline construction and drew widespread media coverage, highlighting the issue to a broader audience (UNICEF, 2023).

Another example is the anti-plastic movement, which has gained significant traction globally. Campaigns against single-use plastics have led to bans and restrictions in several countries. This success is due in large part to collective advocacy efforts involving environmental organizations, consumers, and businesses. By refusing plastic straws and bags, supporting legislative changes, and promoting alternatives, ordinary citizens have made substantial contributions to reducing plastic pollution.

Digital activism plays a critical role in these movements. Online petitions, crowdfunding for legal battles, and creating viral hashtags

(#BreakFreeFromPlastic) are some of the ways activists leverage the internet to drive campaigns. Engaging celebrities and influencers can also amplify messages, reaching even wider audiences and attracting more support.

Staying updated and informed is essential for maximizing the impact of advocacy efforts. Reliable sources such as the Society of Environmental Journalists and the Climate Reality Project provide timely news and insights that can inform your strategies. Continuous learning about climate science, policy developments, and effective advocacy techniques is not only recommended but necessary for successful activism (University of Minnesota, 2023).

Reflecting on these points, it's clear that everyone has a role to play in climate advocacy. Individual actions, when localized and shared, can spark larger movements. Collective action magnifies these efforts, making them more potent. Modern tools for advocacy, particularly social media, offer unparalleled opportunities for outreach and engagement. Finally, looking to successful citizen advocacy movements provides both inspiration and practical lessons for future campaigns.

Final Thoughts

This chapter delved into the significant impact of policy changes and advocacy in combating climate change. By understanding how global treaties like the Kyoto Protocol and Paris Agreement influence national strategies, we can see the critical role these agreements play in setting frameworks for climate action. The collaboration between governments and grassroots movements is essential in driving systemic reforms. International cooperation ensures that nations work together, sharing resources and knowledge to meet ambitious climate goals.

However, challenges such as political opposition and socio-economic disparities can hinder the effectiveness of these policies. It's important to recognize the ongoing efforts of NGOs and other organizations in filling these gaps by advocating for stronger environmental laws and mobilizing public support. Moving forward, fostering these collaborations and addressing equity concerns will be crucial for effective climate action. By setting measurable goals and continuously adapting our approaches, we can build a more sustainable future while ensuring that vulnerable populations are not left behind.

Chapter 8

Indigenous Knowledge: Wisdom from Traditional Ecological Perspectives

Indigenous knowledge encompasses a vast array of insights and practices derived from centuries of close interaction with nature. This chapter explores how such traditional ecological wisdom can be seamlessly integrated into contemporary climate strategies to enhance sustainability. By examining the time-tested methods developed by indigenous communities, we gain valuable perspectives on how to manage our natural resources more responsibly and sustainably.

Throughout the chapter, you will delve into various indigenous agricultural techniques that have proven effective in maintaining soil health and fostering biodiversity. You will learn about milpa systems from Mesoamerica, where crop diversity and soil fertility techniques contribute to ecosystem resilience. The concept of permaculture, inspired by indigenous practices, highlights the importance of mimicking natural ecosystems for sustainable productivity. Additionally, you will discover coastal resource management principles that promote species conservation and community-driven initiatives, as well as agroforestry practices that integrate trees and shrubs into agricultural landscapes for land restoration. By recognizing and valuing these traditional practices, we can create a holistic approach to sustainability that benefits both the environment and human communities.

Traditional Agricultural Practices and Sustainability

Traditional agricultural methods have long demonstrated their value in fostering sustainability and food security, and they remain pertinent as we face modern climate challenges. As such, understanding and integrating these practices can significantly enhance our efforts toward sustainable development.

Milpa systems, an ancient agricultural practice from Mesoamerica, are a prime example of how crop diversity and soil fertility techniques contribute to ecosystem resilience. In a milpa system, multiple crops such as maize, beans, squash, and chili are planted together. This polyculture approach not only maximizes land use but also promotes biodiversity. The different plants complement each other; for instance, beans fix nitrogen in the soil, which benefits maize, while squash provides ground cover that retains soil moisture and suppresses weeds. By emulating natural ecosystems, milpa farmers create a symbiotic environment where pests and diseases are naturally managed, reducing the need for chemical interventions. Additionally, the continuous rotation of these crops helps maintain soil health, preventing nutrient depletion and erosion. Overall, the milpa system exemplifies how traditional agricultural wisdom can build robust and resilient ecosystems (Altieri & Toledo, 2011).

Another critical framework is permaculture, which draws heavily from indigenous practices. Permaculture principles focus on mimicking natural ecosystems to achieve higher productivity sustainably. These principles include creating diverse plant systems that work in harmony, ensuring efficient energy use, and recycling nutrients within the system. For instance, planting perennials alongside annuals creates layered farming structures that use vertical space efficiently and protect soil structure. Permaculture also emphasizes water conservation through techniques like rainwater harvesting and swales, which capture and store water in the landscape. This method reduces food waste by promoting local food production and consumption cycles, directly addressing environmental impacts associated with long-distance food transport. Indigenous communities have long practiced these principles, demonstrating their effectiveness in achieving sustainability and food security (Altieri & Toledo, 2011).

Coastal resource management among indigenous coastal communities offers valuable lessons in sustainable fishing practices. Traditional fishing methods reduce overfishing and promote species conservation through community-driven initiatives. Practices such as seasonal closures and the use of specific tools designed to target only mature fish ensure the continued abundance of marine life. For example, certain Pacific Islander communities employ fish traps that allow younger fish to escape, ensuring that populations remain healthy and reproductive. These methods rely heavily on local knowledge and communal cooperation, reinforcing social cohesion and shared responsibility for resource management. Such practices highlight how indigenous wisdom can guide contemporary efforts to manage natural resources responsibly and sustainably (Sustainability in Mexico - Original Travel, n.d.).

In the realm of restoration agriculture, indigenous strategies like agroforestry play a crucial role in restoring degraded lands. Agroforestry involves integrating trees and shrubs into agricultural landscapes, enhancing biodiversity, and providing ecological benefits such as improved soil structure, increased carbon storage, and better water retention. Trees in agroforestry systems act as windbreaks, reduce soil erosion, and offer habitats for various species, contributing to a thriving ecosystem. Furthermore, these systems often produce diversified yields, including fruits, nuts, and timber, reducing dependency on single-crop farming and increasing resilience against market fluctuations. Indigenous communities have perfected these techniques over centuries, showing us that it is possible to meet human needs while nurturing the environment. Incorporating these practices into modern agriculture can help address pressing issues of land degradation and climate change mitigation (Altieri & Toledo, 2011).

Land Stewardship Techniques

Indigenous land stewardship techniques are paramount in achieving ecological integrity and sustainability. These practices, honed over countless generations, offer invaluable insights into managing ecosystems effectively and holistically. One prominent method is fire management, where indigenous peoples use prescribed burning to control invasive species and enhance native habitats. This practice involves applying controlled and deliberate fires to specific areas, mimicking natural fire cycles that many ecosystems depend on.

Prescribed burns, also known as cultural burns, have been traditionally used to foster biodiversity. By reducing the density of underbrush and removing dead organic matter, these burns prevent large-scale wildfires which can decimate ecosystems. In addition, controlled burns promote the growth of native plant species that thrive after low-intensity fires. This technique underscores community involvement since tribal members typically conduct these burns, ensuring that the knowledge, timing, and methods passed down through generations are respected and applied accurately.

Moreover, the regeneration of landscapes through indigenous-led practices illustrates a profound connection to the land. Regenerative land practices are essential for healing degraded habitats. Through collaborative restoration efforts, indigenous communities employ traditional methods such as reforestation using native species or rehabilitating wetlands. These efforts not only restore the physical landscape but also rebuild ecological networks, promoting species diversity and ecosystem health. The involvement of the entire community in these projects fosters a sense of ownership and responsibility towards sustainable land management.

Water conservation is another critical aspect of indigenous ecological stewardship. Traditional catchment systems, which include creating reservoirs and channels to manage rainfall and runoff, are effective ways to conserve water resources. These systems are designed to capture and store rainwater, gradually releasing it to support agriculture and replenish groundwater supplies. Furthermore, community engagement plays a crucial role in maintaining and protecting watersheds. Indigenous peoples often view water sources as sacred, advocating for their protection and sustainable use to ensure the availability of clean water for future generations.

The management of culturally important species also contributes significantly to biodiversity and ecosystem resilience. Many indigenous cultures hold certain species as sacred or integral to their heritage. These species, whether they are plants, animals, or fungi, are meticulously managed to ensure their survival and proliferation. This management includes monitoring population levels, safeguarding habitats, and sometimes performing rituals or ceremonies to honor these species. By maintaining these species, indigenous communities help sustain the broader ecosystem functions they support, emphasizing the interconnectedness of all life forms.

For instance, the reintroduction of bison in some Native American lands showcases how traditionally important species can rejuvenate ecosystems. Bison grazing patterns help prevent overgrowth of certain grasses, promoting a diverse range of flora that benefits numerous other species. Such examples highlight the crucial role indigenous knowledge plays in fostering ecological balance.

Incorporating these practices into modern environmental strategies can greatly enhance sustainability efforts. Recognizing and valuing indigenous knowledge systems allows for a more holistic approach to conservation, one that integrates scientific understanding with time-tested traditional methods. Policymakers and environmentalists must collaborate with indigenous communities, creating frameworks that respect and utilize these valuable insights.

To illustrate, modern fire management policies can benefit from integrating prescribed burns. Agencies like the National Park Service, which have already begun to incorporate cultural burns in their programs, demonstrate the practical benefits of this approach. Indigenous-led burns help reduce fuel loads, thus preventing catastrophic wildfires, while simultaneously promoting biodiversity by maintaining habitats that many native species rely on. (Hoffman et al., 2021).

Similarly, water conservation initiatives can be significantly enhanced by adopting traditional catchment systems. These systems not only prove efficient in water management but also represent sustainable engineering practices that require fewer resources and less labor compared to contemporary methods. Community-based approaches to watershed management, inspired by indigenous stewardship, can lead to more resilient water resources in the face of climate change.

Lastly, the protection of culturally important species is pivotal in conserving biodiversity. Indigenous-driven conservation programs often identify key species that serve as indicators of ecosystem health. Protecting these species ensures the stability of the entire ecosystem, as each species plays a unique role within it. Efforts to maintain populations of culturally significant species must be supported at both local and national levels, recognizing their broader ecological importance.

Cultural Approaches to Biodiversity Conservation

Indigenous communities have long maintained a profound connection with their natural surroundings, viewing the earth and its ecosystems as integral parts of their cultural identity. This deep-seated relationship manifests in various cultural practices that significantly contribute to biodiversity conservation efforts. This subpoint delves into how these practices function and their importance in modern-day conservation strategies, highlighting the vital role indigenous wisdom plays in sustaining our planet.

Cultural narratives among indigenous communities are more than mere stories; they are powerful tools for fostering ecological ethics and behaviors. These narratives often feature elements of nature as protagonists. Myths, legends, and folklore convey moral lessons emphasizing respect and responsibility towards the environment. For example, within many Native American tribes, tales about animals and plants inculcate values of coexistence and stewardship. These stories serve an educational purpose, instilling in younger generations a sense of duty to protect and preserve their natural heritage. By passing down ecological knowledge through engaging storytelling, these communities cultivate an intrinsic motivation to practice sustainable living, which aligns seamlessly with contemporary conservation goals.

Rituals and seasonal practices also play a critical role in indigenous conservation efforts. Many ceremonies and communal activities are intricately linked to the rhythms of the seasons and the migratory patterns of species. For instance, the Maori people of New Zealand observe rituals such as Matariki, celebrating the Pleiades star cluster's rise, which marks the beginning of the harvest season. These gatherings are not merely celebratory but also educational, reinforcing sustainable harvesting techniques and ensuring that community members adhere to practices that do not deplete resources. Through aligning human activities with natural cycles, these practices help maintain ecological balance and biodiversity.

Central to many indigenous worldviews is the perception of nature as kin. This viewpoint fosters restorative relationships with the environment, promoting sustainable harvesting practices and environmental protection. The concept of "kaitiakitanga" in Maori culture exemplifies this approach, where humans act as guardians of the environment out of reverence for their ancestors and concern for future generations (UN Environment Programme, 2017). This relational perspective encourages practices like rotational farming, selective logging, and regulated hunting, which ensure that ecosystems can regenerate and sustain wildlife populations. Viewing nature as a relative invokes a sense of personal responsibility and care that modern conservation initiatives often strive to emulate.

The integration of cultural approaches into formal conservation plans has shown considerable success through participatory conservation initiatives. Collaborations between indigenous communities and conservation organizations leverage local knowledge and practices to enhance biodiversity outcomes. One illustrative example is the Maasai Wilderness Conservation Trust (MWCT) in Kenya, which works closely with the Maasai people to safeguard Chyulu Hills. The trust employs traditional Maasai practices alongside scientific methods to protect the region's rich biodiversity, including iconic species like elephants and lions (*CEPF*, n.d.). Such initiatives demonstrate that involving indigenous communities in conservation planning and implementation enriches the process with generations of accumulated ecological wisdom and ensures culturally sensitive and effective strategies.

Participatory conservation initiatives also empower indigenous groups, safeguarding their rights while enhancing biodiversity. In Colombia, the InkalAwa people received assistance from CEPF to develop a management plan for their ancestral land, the Pialapí Pueblo Viejo Indigenous Reserve. This project not only equipped the community with skills in reserve management but also fostered a stronger connection between the people and their land (*CEPF*, n.d.). Empowering indigenous communities to take the lead in conservation efforts acknowledges their sovereignty and nurtures their role as custodians of critical ecosystems.

The symbiotic relationship between indigenous peoples and biodiversity underscores the indispensable value of integrating traditional knowledge into contemporary conservation strategies. Cultural narratives reinforce ecological responsibilities, while rituals aligned with natural cycles promote sustainable resource use. The view of nature as kin nurtures restorative practices, and participatory initiatives honor indigenous expertise and leadership in conservation efforts. Recognizing and supporting these cultural practices not only preserves biodiversity but also upholds the rights and traditions of indigenous communities, creating a holistic approach to sustainability that benefits all.

Restoration Agriculture

Indigenous knowledge offers a treasure trove of strategies for restoring degraded lands while enhancing biodiversity and carbon storage. Many indigenous communities have been practicing sustainable land management techniques for generations, demonstrating an inherent understanding of ecological balance and resilience.

One transformative practice employed by many indigenous cultures is agroforestry. This technique involves integrating tree planting with crops, providing numerous benefits to the ecosystem. Trees in agroforestry systems serve multiple functions: they improve soil health by adding organic matter and nutrients, protect against erosion, and offer habitat for a variety of species, thus boosting biodiversity. Additionally, trees act as carbon sinks, sequestering carbon dioxide from the atmosphere and mitigating climate change. Agroforestry is not just about planting trees; it's a holistic approach that considers the symbiotic relationships between flora, fauna, and human activity.

Community-led initiatives are another crucial component of effective land restoration. Indigenous and local communities possess detailed knowledge of their environments, accumulated over centuries. By involving these communities in restoration projects, efforts are more likely to succeed both ecologically and socially. Community participation fosters a sense of ownership and responsibility, ensuring long-term commitment to the preservation and rehabilitation of the land. These initiatives also promote social cohesion, as collective efforts towards a common goal can strengthen communal bonds and shared values. For example, in Nepal, devolving state forests to local community management has significantly slowed deforestation and restored communal forests (S. et al., 2022).

Adopting a nurturing mindset towards land use is another key principle derived from indigenous wisdom. Unlike exploitative practices that deplete resources, a nurturing approach emphasizes care and regeneration.

This mindset encourages us to view the land as a living entity that needs tending and respect. It translates into sustainable practices that prioritize the long-term health of ecosystems over short-term gains. Such an approach aligns perfectly with the concept of stewardship, where humans see themselves as caretakers of the earth rather than its conquerors.

Traditional methods, such as crop rotation and fallowing, are time-tested techniques that remain relevant today. Crop rotation involves alternating different crops in the same area across seasons, which helps maintain soil fertility and disrupt pest cycles. Fallowing, or allowing land to rest by not planting crops for a period, enables soil to recover and rebuild its nutrient profile. Both these methods enhance agricultural sustainability and soil health, ensuring that the land remains productive for future generations. In fact, scientific research has recognized the effectiveness of these age-old practices in modern ecological restoration efforts (Restoration of Degraded Ecosystems Using Traditional Knowledge | Native Peoples and the Environment Class Notes | Fiveable, 2024).

Incorporating indigenous strategies into contemporary climate strategies is not without challenges. Restoration requires continuous effort and adaptation to specific local conditions. Successes in one region may not be directly transferable to another due to differing ecological contexts. However, the adaptability of traditional ecological knowledge makes it highly valuable. Indigenous communities often employ adaptive management strategies based on long-term observations, adjusting practices as environmental conditions change. This flexibility ensures that restoration efforts remain relevant and effective over time.

Furthermore, the spiritual and cultural connections that indigenous peoples have with their land play a vital role in conservation. These ties foster a deep sense of respect and responsibility towards natural resources, encouraging sustainable management practices. For instance, the Guarani people's prohibition of farming in sacred areas ensures the protection of critical freshwater springs and forests, highlighting how cultural beliefs can drive effective environmental stewardship (S. et al., 2022).

Engaging local and indigenous communities in restoration projects also brings additional socio-economic benefits. Restoration activities can create livelihood opportunities, reduce poverty, and enhance food security. When communities are directly involved, they are more likely to develop innovative solutions tailored to their specific needs and circumstances. This bottom-up approach contrasts with top-down policies that may overlook local realities and priorities.

Final Insights

By delving into the rich traditions of indigenous agricultural practices, this chapter illustrates how ancient wisdom can significantly strengthen modern sustainability efforts. We explored various techniques like the milpa system and permaculture principles that enhance soil health, promote biodiversity, and reduce chemical dependency. We also examined the role of indigenous coastal communities in sustainable fishing and how agroforestry aids in restoring degraded lands. These time-tested methods showcase the potential of integrating traditional ecological knowledge into contemporary climate strategies, offering practical solutions to environmental challenges.

Moreover, the chapter highlighted the importance of community involvement and cultural connections in promoting sustainable land stewardship. Practices such as prescribed burns, water conservation systems, and the management of culturally significant species underscore the holistic approach indigenous peoples take towards ecosystem management. By recognizing and incorporating these traditional practices, modern strategies can become more effective and resilient. Embracing indigenous wisdom not only fosters ecological balance but also honors the deep-seated relationships between communities and their natural environments, paving the way for a more sustainable future.

Chapter 9

Innovative Technologies: The Future of Climate Solutions

Innovative technologies are shaping the future of climate solutions, offering new ways to tackle the challenges of global warming and environmental degradation. These advancements are redefining how we produce, store, and utilize energy, making it possible to reduce our dependency on fossil fuels while promoting sustainability. From breakthroughs in renewable energy sources to emerging methods of storage and usage, technology is at the heart of this transformation.

In this chapter, we will explore various pioneering technologies that are revolutionizing the fight against climate change. We'll delve into the latest developments in solar and wind energy, examining how these innovations increase efficiency and make renewable energy more accessible. Furthermore, we will discuss the role of hydrogen fuel as a clean alternative for sectors difficult to electrify, alongside advancements in energy storage that ensure a stable supply even when natural conditions fluctuate. The chapter also highlights the importance of integrating these technologies into everyday life, from residential to industrial applications, underlining their potential to drive us toward a sustainable future.

Renewable Energy Innovations

The future of climate solutions lies in the innovative technologies driving renewable energy advancements. These breakthroughs are crucial as they hold the potential to transform global energy systems, reducing our dependency on fossil fuels and promoting a more sustainable and equitable world.

One significant area of focus is solar energy advancements. Recent developments in solar panel technology have led to the creation of more efficient panels that can capture greater amounts of sunlight. Innovations like bifacial solar panels, which can absorb sunlight from both sides, significantly boost energy capture. Furthermore, advances in solar storage technologies, such as solar batteries, allow for better retention of captured energy. This means that energy generated during sunny periods can be stored and used during non-sunny times, enhancing reliability. Integrating these technologies helps reduce reliance on traditional energy sources and promotes energy equity by making renewable energy accessible to more communities. To maximize the potential of solar energy, it's essential to incorporate solar power into residential, commercial, and industrial applications, ensuring widespread adoption and impact.

Wind power enhancements are another vital component of the renewable energy landscape. Traditional wind turbines have undergone significant transformations with the development of vertical-axis wind turbines (VAWTs). Unlike traditional horizontal-axis turbines, VAWTs can operate efficiently regardless of wind direction, making them suitable for urban environments where wind patterns can be unpredictable. Additionally, offshore wind farms represent a promising avenue for energy generation. By harnessing stronger and more consistent winds over the ocean, these farms can produce substantial amounts of electricity. Innovations in turbine design, such as larger rotor diameters and floating wind turbines, contribute to more efficient energy production. Offshore wind projects also create local jobs and stimulate economic growth in coastal regions. To support these advancements, policies encouraging investment in wind energy infrastructure and technological research are essential.

Hydrogen fuel development is emerging as a clean and versatile solution for energy storage and transportation. Green hydrogen, produced using renewable energy sources, offers a sustainable alternative to traditional fuels. It has the potential to decarbonize sectors that are difficult to electrify, such as heavy industry and long-haul transportation. The European Union's ambitious hydrogen strategy aims to install 40 GW of electrolyzers by 2030, underscoring the growing commitment to this technology. Green hydrogen can be stored and used when needed, providing a reliable energy source even when other renewables are not available. Its production process involves breaking water molecules into hydrogen and oxygen using electrolysis, powered by renewable energy. This approach produces zero emissions, making it an environmentally friendly option. Supporting the development of green hydrogen infrastructure and fostering public-private partnerships can accelerate its adoption and integration into various industries.

Emerging energy storage solutions play a critical role in addressing the intermittency of renewable energy sources like solar and wind. Advanced battery technologies, including solid-state batteries, are at the forefront of this innovation. Solid-state batteries use solid electrolytes instead of liquid ones, offering higher energy density, longer lifespan, and improved safety. These batteries can enhance energy resilience, ensuring a stable supply during peak demand periods. Additionally, grid-scale battery storage projects, such as

Tesla's Megapack installations, provide backup power and grid stability, further bolstering the reliability of renewable energy systems. Efficient and affordable energy storage solutions are essential for maximizing the benefits of renewable energy. As researchers continue to explore new materials and designs for batteries, the barriers to widespread adoption will diminish, leading to a more resilient and sustainable energy grid.

To ensure the successful implementation of these renewable energy technologies, several guidelines should be followed. For solar energy, individuals and businesses should consider investing in high-efficiency solar panels and integrating them with advanced storage systems. Governments and policymakers can support this transition by providing incentives and subsidies for solar installations, as well as funding research and development in solar technology. In the realm of wind power, supporting the construction of VAWTs in urban areas and promoting offshore wind projects through favorable regulations and financial incentives can drive significant progress.

Collaboration between private companies, government agencies, and research institutions is crucial for advancing wind energy technologies.

For hydrogen fuel development, creating a robust infrastructure for hydrogen production, storage, and distribution is essential. Governments should foster international cooperation and establish standards and regulations for the safe handling and use of hydrogen. Public awareness campaigns can also help educate communities about the benefits of green hydrogen and its role in achieving a sustainable future. Finally, for emerging energy storage solutions, continued investment in battery research and development is paramount. Supporting pilot projects and large-scale deployments of advanced storage technologies can demonstrate their viability and encourage broader adoption.

Carbon Capture and Storage Technologies

Carbon capture and storage (CCS) technologies have emerged as a pivotal strategy in the fight against climate change. These innovations play a crucial role in reducing CO_2 levels in the atmosphere, supporting efforts to mitigate global warming and its adverse impacts. As we delve into this topic, it's essential to understand the various facets of CCS and their implications for our sustainable future.

One of the most promising aspects of CCS is Direct Air Capture (DAC). This technology has garnered significant attention due to its potential to directly remove CO_2 from the atmosphere. Unlike traditional methods that only target emissions at their source, DAC works by filtering ambient air to extract CO_2, which can then be stored or utilized. Companies like Climeworks and Carbon Engineering are at the forefront of this innovation, developing systems that can capture thousands of tons of CO_2 annually. While DAC is still relatively new and costly, it presents an exciting avenue for achieving net-zero emissions targets. Moreover, the captured carbon can be repurposed in various industries, fostering the growth of new markets centered around carbon utilization.

Transitioning from atmospheric capture, Point Source Capture Innovations focus on intercepting emissions from specific sources like power plants and industrial facilities. These advancements are critical in curbing emissions from some of the largest contributors to atmospheric CO_2. Techniques such as post-combustion capture, pre-combustion capture, and oxy-fuel combustion have seen significant improvements in efficiency and cost-effectiveness. By refining these technologies, it becomes more feasible for industries to adopt CCS solutions without facing prohibitively high operational costs. For instance, companies can retrofit existing facilities with CCS equipment, substantially reducing their carbon footprint while contributing to a circular economy. By capturing CO_2 directly at the emission point, we not only reduce atmospheric pollution but also pave the way for reutilizing carbon in industrial processes, thus promoting sustainability.

Equally important are the advances in Geologic Storage, which ensure the safe and long-term containment of captured CO_2. This method involves injecting CO_2 into underground geological formations like depleted oil fields or deep saline aquifers. To facilitate broader deployment of CCS, it is crucial to enhance the safety and reliability of these storage methods. Research and development in this area have led to improved monitoring techniques, like seismic imaging and pressure monitoring, which track the movement of CO_2 and ensure it remains securely stored. Additionally, regulatory frameworks and best practices have been established to minimize environmental risks and guarantee maximum safety. Countries like Norway have successfully implemented large-scale geologic storage projects, demonstrating the viability and effectiveness of this technique. The success of these initiatives provides valuable insights and models that other regions can adopt, making geologic storage a mainstay of CCS strategies globally.

To accelerate the adoption of CCS technologies, innovative financing mechanisms are essential. Funding these projects often requires substantial investment, which can be a barrier for widespread implementation. However, new strategies are being developed to attract private investment and create financial incentives. Public-private partnerships have proven particularly effective, combining government support with private sector expertise and resources. These collaborations can drive research and development, facilitating the commercialization of CCS technologies. Furthermore, market-based mechanisms like carbon pricing and tax credits can incentivize companies to invest in CCS, making it a competitive option alongside traditional practices. For example, the 45Q tax credit in the United States provides financial benefits to businesses that capture and store CO_2, encouraging the adoption of these technologies. By creating a favorable economic environment for CCS, we can ensure that these solutions are not only technically feasible but also financially viable for widespread use.

Smart Agriculture and Resource Management Systems

Innovations aimed at optimizing agricultural practices and resource management hold immense potential for enhancing sustainability and reducing environmental impact. This section delves into several key advancements in agricultural technology that promise to reshape the future of farming.

Precision Agriculture is a revolutionary approach that leverages big data and Internet of Things (IoT) devices to improve farming efficiency and reduce the use of resources like water, fertilizers, and pesticides. By collecting and analyzing vast amounts of data on soil conditions, weather patterns, and crop health, precision agriculture enables farmers to make informed decisions on when and how to apply inputs. For example, sensors placed in fields can monitor soil moisture levels, enabling irrigation systems to deliver the exact amount of water needed, thus conserving this vital resource. Similarly, drones equipped with imaging technology can identify areas of pest infestation early, allowing targeted pesticide application instead of blanket spraying. These technologies not only enhance productivity but also contribute to healthier ecosystems by minimizing chemical runoff and waste (Mass Challenge, 2022).

Vertical Farming Innovations represent another significant advancement in sustainable agriculture. By growing crops in vertically stacked layers within controlled environments, urban and indoor farming maximize space utilization and significantly reduce the carbon footprint associated with transporting produce over long distances. Vertical farms can be established in urban centers, providing fresh food to city dwellers while integrating sustainable practices such as aquaponics, where crops and fish are farmed together in a symbiotic environment. The closed-loop systems used in vertical farming recycle water and nutrients, making it a highly efficient method of production. Additionally, LED lighting tailored to the specific needs of different plants ensures optimal growth conditions year-round, independent of external climate variability. These innovations not only enhance local food security but also mitigate the urban heat island effect by incorporating green spaces into cities.

Agroecological Approaches involve integrating ecological principles into agricultural practices to restore and maintain biodiversity, improve soil health, and reduce reliance on chemical fertilizers. This approach encourages diversity in cropping systems and the adoption of organic farming techniques. Practices such as crop rotation, intercropping, and the use of cover crops help maintain soil fertility and structure, reducing erosion and improving water retention. Agroforestry, which combines trees and shrubs with crops or livestock, provides multiple ecosystem services including carbon sequestration, habitat for wildlife, and improved resilience to climatic fluctuations. By promoting a diverse array of plant species, agroecological approaches enhance pest management naturally and foster more stable agricultural systems.

Resource Management Systems play a crucial role in advancing sustainable agriculture by utilizing smart technologies for better management of water and soil health. Smart water management systems integrate sensors and automated controls to optimize irrigation schedules based on real-time data. This prevents over-watering and conserves water, which is especially critical in regions experiencing water scarcity. Soil health monitoring technologies provide detailed insights into soil composition and nutrient levels, enabling precise application of fertilizers only where needed, thus preventing nutrient runoff and groundwater contamination. For instance, the use of satellite imagery and remote sensing technology allows farmers to assess large areas of land efficiently and determine the specific needs of each section, facilitating site-specific management practices. Such technologies promote sustainable land management by ensuring that resources are used judiciously, leading to long-term soil health and productivity.

The transformative potential of these agricultural innovations lies not just in their technical sophistication but in their ability to address the pressing need for sustainable food production in the face of climate change and growing global populations. Precision agriculture, with its data-driven insights, empowers farmers to fine-tune their practices to achieve maximum efficiency and minimal environmental impact. By reducing the dependency on excessive inputs, it fosters an agricultural system that is both economically and ecologically viable.

Vertical farming tackles the challenge of feeding urban populations while reducing transportation emissions and the environmental impacts of conventional farming. It demonstrates how innovative designs and localized food production can effectively contribute to urban sustainability and resilience. The integration of advanced technologies in vertical farming setups showcases a blend of modern engineering and ecological mindfulness, setting a precedent for future agricultural practices.

Agroecological approaches underline the importance of working with nature rather than against it. By nurturing biodiversity and soil health, these methods create robust agricultural systems capable of withstanding environmental stresses. They highlight the interconnectedness of farming and natural ecosystems, advocating for practices that sustain both.

Resource management systems epitomize the application of technology to resource conservation. In an era where water scarcity and soil degradation are major concerns, such systems offer practical solutions to manage these critical resources intelligently. By aligning agricultural practices with environmental stewardship, they pave the way for sustainable land use that supports both current and future generations.

Emerging Technologies and Innovative Materials

Groundbreaking materials and emerging technologies are pivotal in advancing sustainability and addressing climate change. In this section, we will explore four key innovations: advanced biofuels, carbon nanotubes and graphene, smart grids, and biodegradable and sustainable materials.

Advanced biofuels represent a significant leap forward in the quest for sustainable energy sources. These biofuels are derived from waste materials such as agricultural residue, municipal waste, and non-food crops, making them a sustainable alternative to traditional fossil fuels. By repurposing waste, advanced biofuels can significantly reduce greenhouse gas emissions that result from transportation and industrial activities. For instance, bioethanol and biodiesel, produced from plant materials, offer cleaner combustion compared to conventional petroleum-based fuels (Bradu et al., 2022). Moreover, advanced biofuels contribute to a circular economy by creating value from waste products, thereby reducing landfill usage and mitigating pollution.

Carbon nanotubes and graphene are two groundbreaking materials with far-reaching implications for various industries. Carbon nanotubes are cylindrical structures made of carbon atoms, renowned for their exceptional electrical, thermal, and mechanical properties. Graphene, a single layer of carbon atoms arranged in a two-dimensional honeycomb lattice, is celebrated for its strength, flexibility, and conductivity. These materials are enhancing performance and reducing environmental impact across multiple sectors. For example, in energy storage, carbon nanotubes and graphene are being utilized to develop more efficient batteries with higher capacity and faster charging times. In electronics, their superior conductivity can lead to smaller, faster, and more energy-efficient devices. The construction industry is also benefiting, as these materials are used to create stronger and lighter building components, which can withstand extreme conditions and have a longer lifespan (Bradu et al., 2022). By incorporating these advanced materials, industries can achieve greater efficiency and sustainability.

The implementation of smart grid technologies is another transformative development in the realm of energy management. Smart grids enable real-time monitoring and control of energy flow, allowing utilities to optimize energy distribution, enhance reliability, and better integrate renewable energy sources. Traditional power grids often struggle with inefficiencies, such as energy loss during transmission and an inability to balance supply and demand dynamically. In contrast, smart grids employ advanced sensors, communication networks, and data analytics to address these challenges. They facilitate demand response programs where consumers can adjust their energy use during peak times in exchange for incentives, thus reducing strain on the grid and promoting energy conservation. Additionally, smart grids support the integration of renewable energy sources like solar and wind by efficiently managing the intermittent nature of these energy forms. For instance, when excess solar energy is generated during the day, it can be stored or redistributed to areas with higher demand, ensuring a balanced and reliable energy supply. Overall, smart grids contribute to a more resilient and sustainable energy infrastructure.

Biodegradable and sustainable materials are essential in reducing pollution and fostering a circular economy. Traditional plastics and packaging materials are major contributors to environmental pollution, taking hundreds of years to decompose. Innovations in biodegradable plastics, made from natural sources such as cornstarch, cellulose, and other plant-based materials, are providing viable alternatives. These materials break down more quickly and with less environmental impact, helping to alleviate the burden of plastic waste in landfills and oceans. For example, polylactic acid (PLA) is a renewable and biodegradable polymer derived from fermented plant starch, commonly used in packaging, disposable cutlery, and medical implants. Besides their environmental benefits, biodegradable materials can enhance the economic viability of recycling programs by simplifying waste separation and processing.

Sustainable packaging solutions also include designs that minimize material use while maximizing functionality and lifespan. This approach not only reduces waste but also conserves resources and energy required for production. Furthermore, companies are exploring innovative methods to recycle and upcycle materials, transforming waste into valuable products. For instance, some manufacturers are using post-consumer waste, such as plastic bottles and old textiles, to create new fabrics and construction materials. By closing the loop of material use, these practices align with the principles of a circular economy, where resources are continuously reused and recycled, minimizing environmental impact.

Concluding Thoughts

As we've explored, the technological advancements in renewable energy are paving the way for a more sustainable future. Solar and wind energy innovations are making clean energy more efficient and accessible, while green hydrogen and advanced energy storage solutions offer reliable alternatives to traditional fossil fuels. These breakthroughs not only reduce our carbon footprint but also promote energy equity and stimulate economic growth. By integrating these technologies into various sectors, we can significantly decrease our reliance on non-renewable resources and move towards a cleaner, greener world.

The journey towards sustainability requires collective effort and supportive policies to ensure these technologies reach their full potential. Governments and private entities must work together to invest in research, infrastructure, and public awareness campaigns to drive adoption. From solar panels in residential areas to offshore wind farms and hydrogen fuel projects, every step counts. Embracing these innovations allows us to address climate change proactively and create a future where renewable energy is the norm, benefiting both current and future generations.

Chapter 10

Optimal Locations in America for Climate Change Resilience

Finding the best places in America for climate change resilience involves more than just checking the weather forecast. It's about diving deep into a blend of geography, infrastructure, and innovation to determine where communities can thrive amid changing climates. As concerns over rising sea levels, temperature shifts, and extreme weather events grow, so does the importance of identifying cities and states that are not only prepared but also equipped to adapt and flourish. These locations don't just offer a safe haven; they provide a model of sustainability and forward-thinking policies that other regions can learn from. By pinpointing these areas, individuals and families can make informed decisions about where to live, work, and invest, ensuring their futures are safeguarded against environmental upheaval.

This chapter explores cities and states across the U.S. with remarkable climate resilience strategies. From analyzing historical climate data to understanding the impact of natural resources and local policies, we'll delve into what makes certain regions stand out in terms of adaptation and sustainable development. You'll discover how different areas leverage technology, infrastructure, and community engagement to combat climate challenges. We'll look at the role of natural resources like forests and water bodies in mitigating harsh weather effects and examine cities leading the charge in proactive environmental policies. Through case studies, readers will gain insights into innovative practices that enhance resilience, ensuring these communities not only survive but thrive as the climate continues to change. This exploration serves as both a guide for potential relocators and a beacon for other cities seeking adaptable solutions in facing future climate uncertainties.

Understanding Future Climate Threats

Identifying regions in America where climate change poses minimal threats involves a systematic assessment of various environmental factors. It's key to first consider the regions vulnerable to extreme weather occurrences such as hurricanes, droughts, and wildfires. States like Florida and Louisiana frequently face hurricanes due to their geographical location near warm ocean waters. This makes their coastal cities highly susceptible to intense storms and flooding. On the other hand, California and parts of the western United States experience severe droughts and wildfires, putting immense stress on water resources and agriculture. To predict these patterns efficiently, we need to examine historical records and trends in weather data.

Next is understanding the implications of sea level rise, particularly for coastal communities. Rising seas threaten to inundate low-lying islands and coastlines, worsening erosion, flooding, and saltwater intrusion that can compromise freshwater resources. Cities like Miami are already experiencing the effects, with "sunny day" flooding becoming more common. Infrastructure in these areas must be adapted urgently to shore up defenses against incoming waters and protect human habitation.

Evaluating historical climate data also plays a critical role in predicting future resilience. By studying past weather records, scientists identify trends that indicate how climates have evolved over time. For example, analyzing temperature rise, precipitation changes, and atmospheric conditions helps paint a picture of what future climates might look like in specific locales. A region's history of surviving previous climate impacts informs its potential to adapt and thrive under changing circumstances.

Infrastructure and technology adaptation potential provides another lens through which to determine a region's capacity to withstand climatic shifts. Investing in climate-smart infrastructure, such as flood-resistant buildings and energy-efficient systems, enhances community resilience. Cities embracing green technologies, like renewable energy sources or electric transportation networks, set a precedent for reducing greenhouse gas emissions and mitigating climate impacts—a forward-thinking step other regions can emulate.

Potential for technology advancement goes beyond physical structures, extending to policies and practices promoting sustainability. Case studies reveal cities with high technology adoption rates often outperform others in resilience metrics. Governments prioritizing innovation around climate solutions attract industries that develop cutting-edge tools for environmental management. This not only boosts regional economies but fosters a culture of sustainable development.

For individuals and businesses considering relocation, understanding these climate challenges and opportunities is paramount. They should assess geographic vulnerabilities, potential for adaptive infrastructure, and local commitment to sustainability initiatives. Regions demonstrating robust climate preparedness, solid policy frameworks, and technological advancements serve as more advantageous options for long-term settlement.

Cities and States with Favorable Climates

When contemplating climate resilience in U.S. cities, it's essential to highlight those with mild temperature fluctuations and stable weather patterns. Cities like San Diego, California, boast a Mediterranean climate characterized by temperate summers and mild winters. Such stability reduces the strain on infrastructure and public health systems, as extreme temperatures often lead to increased energy demands and heat-related illnesses. Similarly, Portland, Oregon, enjoys moderate weather conditions, which can significantly enhance livability while minimizing disruptions caused by severe weather events.

Examining states with natural resources that inherently mitigate harsh climate conditions further underscores their suitability for future habitation. For instance, Vermont's abundant forests play a crucial role in absorbing carbon dioxide and providing natural cooling through evapotranspiration. This natural resource is vital for moderating local climates and reducing exposure to high temperatures. Additionally, the presence of numerous lakes and rivers throughout Minnesota offers not only scenic beauty but also acts as a buffer against heatwaves, helping to maintain cooler temperatures.

Locales known for proactive environmental policies and initiatives are another essential consideration when assessing optimal living environments. Boulder, Colorado, has long been at the forefront of climate action, implementing comprehensive sustainability plans that focus on renewable energy, emissions reductions, and sustainable urban planning. These initiatives are crucial for fostering a community that adapts well to climate challenges and promotes long-term resilience. Similarly, Austin, Texas, has made significant strides with its Climate Protection Plan, aiming to reduce greenhouse gas emissions and improve energy efficiency through city-wide programs.

Particular communities have also demonstrated a strong commitment to investing in sustainable practices and technologies, leading the way in creating eco-friendly living spaces. One exemplary model is the city of Ithaca, New York, which has embraced renewable energy solutions, such as solar and wind power, to meet its energy needs sustainably. The community's emphasis on green building standards and transportation improvements aligns with broader efforts to reduce the carbon footprint and enhance environmental quality. Another notable example is the town of Burlington, Vermont, which became the first city in the U.S. to run entirely on renewable energy sources. Burlington's success exemplifies how prioritizing sustainability can effectively combat climate change impacts while fostering economic growth and community well-being.

In selecting climate-safe states, distinct criteria should guide decision-making processes. Favorable locations often exhibit diverse ecosystems capable of adapting to climatic changes, such as varied topographies that offer natural protection against extreme weather. States like Washington benefit from both mountainous regions and coastal areas, providing options for different climate preferences while mitigating risks associated with sea level rise or flooding. Furthermore, evaluating access to clean water resources and fertile land becomes paramount, as these elements are vital for sustaining agricultural activities and ensuring food security amid shifting climate conditions.

The role of altitude in mitigating climate impacts is another critical factor worth considering. Regions situated at higher elevations, such as Denver, Colorado, enjoy cooler temperatures due to their altitude, making them less susceptible to the heat island effect prevalent in lower-lying urban centers. Higher altitudes also contribute to decreased exposure to air pollutants, promoting improved air quality and public health outcomes. Moreover, the natural topography offers opportunities for hydroelectric power generation, enhancing energy security while reducing reliance on fossil fuels.

As we continue to explore optimal locations for future livability, it becomes increasingly evident that a multifaceted approach is necessary. Balancing environmental, social, and economic factors plays a crucial role in determining cities and states best suited for enduring climate change impacts. By prioritizing regions with favorable weather patterns, robust natural resources, progressive policies, and sustainable practices, communities can build resilience and foster adaptability in the face of an uncertain future.

Case Studies: Examples of Resilient Areas

In the face of advancing climate change, certain cities in the United States are becoming exemplars for resilience and adaptation. These cities demonstrate how thoughtful, coordinated efforts can yield tangible benefits in combating climate challenges. By examining their measures, partnerships, and community-driven successes, we can glean valuable insights into effective urban resilience strategies.

One prime example is Washington DC, where a holistic approach to climate adaption has been methodically pursued since the Derecho storm in 2012. Their Climate Ready DC strategy showcases how integrating technical expertise with community resources can create robust adaptation plans. This comprehensive plan targets utilities, infrastructure, buildings, neighborhoods, and governance with goals such as reducing greenhouse gas emissions by 50% by 2032 and improving building designs to withstand climate impacts. Moreover, they've planted over 14,000 trees and installed green roofs, benefiting both environmental and economic sectors through enhancements like increased urban greenery and reduced energy costs (silviapellegrino, 2023).

Copenhagen provides another success story with its proactive Climate Adaptation Plan initiated after severe flooding in 2011. The city's response included systematic flood management through wastewater and biodiversity strategies, ensuring rapid recovery and prevention of future vulnerabilities. Initiatives like increasing urban green spaces and creating networks of biodiversity have enhanced livability while providing nature-based solutions to climate impacts. Copenhagen's dedication to sustainability and resilience places it among the world's most forward-thinking cities regarding climate policy (silviapellegrino, 2023).

Sydney exemplifies innovation in mitigating urban heat island effects through creative adaptations. By repaving darker surfaces with lighter colors and implementing significant tree canopy expansions, Sydney has successfully reduced local temperatures and improved air quality. Such initiatives reflect how localized interventions, supported by citizen panels and democratic decision-making, can foster resilient urban environments. Through water recycling schemes and retrofitting major water users, Sydney demonstrates a commitment to water conservation and urban cooling, contributing to its enhanced livability index (silviapellegrino, 2023).

The partnerships between these cities and various environmental organizations play a vital role in their success. For instance, local governments often collaborate with experts from academic institutions and non-governmental organizations to develop targeted strategies that address specific ecological issues. In Washington DC, for example, collaboration with statistical and technical experts allowed for precise climate projection models and vulnerability assessments, leading to more effective, data-driven adaptation actions (silviapellegrino, 2023).

These collaborative efforts are further strengthened by statistics that underline improvements in livability and sustainability. Data from Sydney highlights a 23% increase in urban tree canopy coverage, directly correlating with better air quality and reduced urban heat. Similarly, Copenhagen's adaptation strategies have led to significantly reduced insurance payouts following extreme weather events. Such statistics not only validate the effectiveness of these measures but also encourage other cities to adopt similar approaches (silviapellegrino, 2023).

Community engagement and innovation are integral to these successes. Local participation in resilience planning ensures that diverse viewpoints shape practical solutions tailored to community needs. In Sydney, public involvement in decision-making processes helped identify key areas for intervention, fostering a sense of ownership and accountability. Washington DC's educational programs empower residents with knowledge on climate risks, enabling informed personal and collective action towards resilience.

Innovation in these cities also manifests through the development of cutting-edge infrastructure and technology. In Copenhagen, sustainable architecture like CopenHill and Klimakvarter housing underscores the merging of aesthetic urban design with functional climate resilience. This integration of green roofs and flood-proofing strategies redefines urban living spaces while safeguarding them against future climatic threats.

A crucial takeaway from these examples is the importance of long-term, adaptable plans focussed on sustainability and resilience. These cities illustrate that facing climate challenges requires continuous innovation, community partnership, and a firm commitment to integrate climate solutions in urban planning. As ecosystems adapt, so must city planners, policymakers, and residents who live within them.

Strategic Planning for Future Settlements

Choosing the right place to settle in the context of climate change is a crucial decision that requires a strategic approach. With increasing awareness and concern about climate impacts, it's essential to evaluate potential settlement areas based on their ability to withstand future climate change challenges. This involves considering various criteria such as natural disaster risk, environmental policies, and community initiatives.

Firstly, understanding what makes a location resilient to climate change is fundamental. Climate resilience can be defined by factors like low susceptibility to natural disasters such as hurricanes, floods, droughts, and wildfires. Areas that have historically experienced fewer extreme weather events are generally more desirable. Additionally, accessibility to abundant freshwater resources, fertile soil, and sustainable energy options can significantly enhance a region's resilience. Analyzing climate models and forecasts can help predict areas less likely to face severe changes in temperature and precipitation patterns, making them safer bets for long-term habitation.

It is also vital for individuals to assess the potential impacts of climate change on personal property and investments. Homebuyers should consider properties built with sustainable materials and designs that minimize environmental impact. For instance, homes that incorporate solar panels, efficient insulation, and rainwater collection systems not only reduce carbon footprints but also offer protection against power outages and water shortages. Property evaluations should include a thorough investigation of the landscape to determine flood zones or areas prone to erosion. Proximity to green spaces and biodiversity can serve as additional buffers against climate-related disruptions, enhancing property value over time.

Long-term urban planning plays an integral role in ensuring cities are prepared for the effects of climate change. Urban planners and policymakers must integrate climate research and adaptation strategies into their development plans. Creating infrastructures that can withstand heavy rainfall or provide cooling in heatwaves is critical. For example, implementing green roofs, permeable pavements, and extensive tree cover can mitigate urban heat island effects and manage stormwater effectively. The collaboration between city planners and climate scientists can generate innovative solutions, transforming cities into resilient hubs.

Moreover, public awareness and involvement in climate-adapted living practices are crucial. Communities that actively participate in climate resilience initiatives are often more successful in implementing effective adaptation measures. Public education campaigns can highlight the importance of reducing individual carbon footprints through sustainable transportation, waste reduction, and energy conservation. Encouraging citizens to partake in community gardens, local conservation projects, and clean-up drives fosters a collective sense of responsibility toward maintaining a sustainable environment. Local governments can enhance engagement by providing incentives for households and businesses that adopt eco-friendly practices.

Developing a personal relocation strategy is another essential guideline when considering moving to an area with better climate resilience. Individuals should start by researching regions that align with their lifestyle preferences while offering a safe environment in terms of climate stability. Access to healthcare, education, and employment opportunities should also factor into these decisions. Additionally, investigating the region's economic health, governance quality, and infrastructure readiness provides a comprehensive view of its suitability. Collaborating with real estate experts who specialize in resilient locations can provide valuable insights and guidance.

Assessing local government initiatives is equally important. Governments that proactively address climate change through legislation and community programs contribute significantly to the region's overall resilience. Reviewing municipal plans for climate adaptation, including investment in renewable energy, sustainable public transport, and emergency preparedness, can reveal how well-prepared a locality is for future challenges. Engaging with local governance structures and community councils can provide firsthand information about ongoing and future projects, helping to gauge commitment levels toward creating climate-resilient communities.

Involvement in community planning is another actionable step individuals can take towards fostering resilience. Participating in neighborhood meetings, supporting local environmental organizations, and voicing opinions during public consultations can drive positive change. As residents engage with each other and authorities, they contribute to a democratic process that shapes the community's long-term vision. Volunteering with local groups focused on sustainability, such as those promoting solar energy installations or advocating for pedestrian-friendly streets, strengthens communal ties and increases resilience.

Finally, investing in sustainable housing and energy solutions is a proactive approach to securing a stable future. Beyond choosing energy-efficient home features, individuals can invest in larger-scale projects that promote sustainability, such as community solar farms or wind energy cooperatives. Such investments not only support local economies but also yield monetary returns while aiding the transition to cleaner energy sources. Homes powered by renewables become less vulnerable to fossil fuel price fluctuations, providing a more secure living environment.

Bringing It All Together

As we conclude this exploration of climate resilience across U.S. cities and states, it's clear that understanding geographic vulnerabilities and adopting innovative strategies are key. Cities like San Diego and Portland show us the benefits of stable weather patterns, while places like Vermont and Minnesota highlight how rich natural resources provide essential buffers against climate extremes. The lessons from proactive communities such as Boulder and Austin emphasize the importance of forward-thinking policies and sustainable practices. These examples underscore the need to evaluate climate resilience through a blend of environmental stability, infrastructure adaptability, and strong policy frameworks.

Taking these insights, individuals and communities can make informed decisions about future settlements in light of climate change. Factors like mild climates, abundant natural resources, and a commitment to sustainability emerge as crucial determinants for long-term livability. As you consider relocation or plan urban developments, these attributes guide where to focus attention and resources. Ultimately, prioritizing areas with robust defenses against climate impacts ensures not just survival, but thriving environments for future generations. By aligning personal and communal goals with these criteria, a more sustainable and resilient future becomes attainable.

Conclusion

Throughout this journey, we've uncovered the multifaceted nature of climate change—be it the alarming realities of rising sea levels or the actionable steps we can each take towards sustainable living. As outlined in Chapter 5, simple lifestyle adjustments can collectively enhance our planet's health. From understanding the science behind climate change to exploring indigenous approaches to conservation, we've delved into areas that shed light on both the problem and the solutions at hand.

We've seen how various sectors are affected by climate change—from agriculture to urban infrastructure—and the ripple effects these changes have on our daily lives. The evidence is clear: our environment is under threat, and the stakes couldn't be higher. Yet, with every chapter, there has been a consistent thread of hope, emphasizing that we hold the power to influence positive change.

It's crucial to remember that every piece of information shared in this book points towards one essential truth: action is needed, and it's needed now. As you reflect on the knowledge gained, remember that every small action contributes to a larger systemic change. Whether it's advocating for community resilience or adopting energy-efficient practices, your choices matter. Together, we can redefine the trajectory of our future.

One of the core messages throughout this book is that while individual actions may seem insignificant in isolation, they accumulate to create substantial impact. Think of the ripple effect—a single drop creating waves that spread far beyond its initial point of contact. This analogy perfectly encapsulates the ethos of sustainability. By making conscious decisions— from reducing plastic use to supporting renewable energy—we collectively steer the world towards a more sustainable future.

Action extends beyond personal habits; it involves engaging with broader initiatives and movements dedicated to combating climate change. We discussed the importance of political advocacy, encouraging readers to support policies that aim for environmental justice and sustainability. Voting for leaders who prioritize green policies, participating in local government meetings, and even grassroots activism can spur significant change. In essence, your voice is a powerful tool, and when used alongside others, it can drive formidable transformation.

Furthermore, education plays a pivotal role in sustaining momentum against climate change. The journey doesn't end with reading this book; it's merely a stepping stone. I urge you to dive deeper into the wealth of resources mentioned throughout this book. Stay informed, attend local meetings, and connect with like-minded individuals. Change is not only possible but requires our collective wisdom and commitment.

Ecological consciousness must be continuously nurtured. To keep growing in your understanding, consider enrolling in courses, joining webinars, and reading up-to-date research articles. Knowledge is power, and by arming yourself with information, you're better equipped to make impactful decisions and inspire those around you to do the same.

Community engagement is another cornerstone of sustained action. The sense of belonging to a collective mission can be incredibly empowering. Join or form local environmental groups, participate in community clean-up events, or start conversations about sustainability in your neighborhood. By fostering local networks, you contribute to a groundswell of environmental stewardship that radiates outward.

Though the challenges are immense, the examples of innovation and advocacy shared in this book illuminate a promising path forward. Imagine a world where renewable energy powers our homes and indigenous wisdom guides sustainable practices—a world within our reach if we work together. Technological advancements continue to offer new tools and methods for reducing our carbon footprint. From solar panels to electric vehicles, technology serves as an ally in our quest for sustainability.

Policies and regulations also play a vital role. By championing policy changes that align with environmental goals, we ensure that efforts to combat climate change are supported by robust legal frameworks. Joining forces with organizations and campaigns that lobby for such policies can amplify your impact. Collective action magnifies individual efforts, pushing society towards more stringent and effective environmental standards.

Moreover, tapping into indigenous knowledge presents an invaluable opportunity. Indigenous communities embody centuries of symbiotic relationships with the earth, their practices offering profound insights into natural resource management and conservation. By embracing and integrating these time-tested methods, we pave the way for sustainable living that honors both tradition and progress.

While it's easy to feel overwhelmed by the magnitude of climate change, focusing on the potential for positive outcomes fuels hope and determination. Case studies, success stories, and visible impacts validate the effectiveness of our efforts. Real-world examples of reforestation projects, ocean clean-ups, and zero-waste communities demonstrate that positive change is not just theoretical—it's happening right now.

Take inspiration from these stories and let them guide your actions. Embrace optimism, not naivety, and understand that setbacks are part of any significant endeavor. Resilience and perseverance will see us through challenges as long as we remain committed to the cause.

In closing, I extend my heartfelt thanks for embarking on this journey of discovery and action. By choosing to educate yourself about climate change and seeking ways to mitigate its impact, you've taken a commendable step toward ensuring a healthier planet. Remember, the fight against climate change is a marathon, not a sprint. Each stride we take, no matter how small, brings us closer to a sustainable future.

So, plant that tree, reduce your waste, vote wisely, and above all, stay curious and motivated. Our planet's future rests in our hands, and together, we can shape a legacy of environmental stewardship that future generations will thank us for. Let's continue this journey, hand in hand, towards a greener, brighter tomorrow.

References

Works Cited

2019). New America. "Climate Migration's Impact on Housing Security in the United States: Recommendations for Receiving Communities." *New America* , 2019, www.newamerica.org/future-land-housing/ reports/climate-migrations-impact-on-housing-security/climate-migration-is-an-opportunity-for-resilience-and-growth/.

2022). Usda.gov. "USDA Opens People's Garden Initiative to Gardens Nationwide." *Usda.gov* , 2022, fsa.usda.gov/news-room/ news-releases/2022/usda-opens-peoples-garden-initiative-to-gardens-nationwide.

"2024). Nrel.gov. ." *Nrel.gov* , 2024, www.nrel.gov/news/program/2024/25-new-coastal-remote-and-island-communities-join-energy-transitions-initiative-partnership-project.html.

Admin. "How Green Energy Innovations Are Shaping a Sustainable World." *Vector Globe* ,

30 Aug. 2024, vectorglobe.com/green-energy-innovations-are-shaping-a-sustainable-world/.

Altieri, Miguel A., and Victor Manuel Toledo. "The Agroecological Revolution in Latin America: Rescuing Nature, Ensuring Food Sovereignty and Empowering Peasants." *Journal of Peasant Studies* , vol. 38, no. 3, July 2011, pp. 587–612, https://doi.org/10.1080/03066150.2011.582947.

Berardi, Mary Kate, et al. "A Community Approach to Disaster Preparedness and Response." *Psu.edu* , 2022, extension.psu.edu/a-community-approach-to-disaster-preparedness-and-response.

Blackburn, Rebecca, et al. "Could a Minimalist Lifestyle Reduce Carbon Emissions and Improve Wellbeing? A Review of Minimalism and Other Low Consumption Lifestyles." *WIREs Climate Change* , vol. 2023, November 11, no. 2020, September, 11 Nov. 2023, https://doi.org/10.1002/wcc.865.

Bradley Patrick White, et al. "Mental Health Impacts of Climate Change among Vulnerable Populations Globally: An Integrative Review."

Annals of Global Health , vol. 89, no. 1, 6 Oct. 2023, pp. 66–66, https://doi.org/10.5334/aogh.4105.

Bradu, Pragya, et al. "Recent Advances in Green Technology and Industrial Revolution 4.0 for a Sustainable Future." *Environmental Science and Pollution Research International* , vol. 30, no. (2022, April 9)., 9 Apr. 2022, pubmed.ncbi.nlm.nih.gov/35397034/, https://doi.org/10.1007/s11356-022-20024-4.

Center for Climate and Energy Solutions. "Hurricanes and Climate Change." *Center for Climate and Energy Solutions* , 26 Sept. 2018, www.c2es.org/content/hurricanes-and-climate-change/.

CEPF. (n.d.). Www.cepf.net. "| CEPF." *Www.cepf.net* , www.cepf.net/stories/indigenous-peoples-and-biodiversity-symbiotic-relationship.

Colson-Fearon, Brionna, and H. Shellae Versey. "Urban Agriculture as a Means to Food Sovereignty? A Case Study of Baltimore City Residents." *International Journal of Environmental Research and Public Health* , vol. 19, no. 19, 5 Oct. 2022, p. 12752, https://doi.org/10.3390/ijerph191912752.

Coolcalifornia.arb.ca.gov. "Sustainable Transportation | Cool California." *Coolcalifornia.arb.ca.gov* , coolcalifornia.arb.ca.gov/sustainable_transportation.

Diekmann, Lucy O., et al. "More than Food: The Social Benefits of Localized Urban Food Systems." *Frontiers in Sustainable Food Systems* , vol. 4, no. Diekmann, L. O., Gray, L. C., & Thai, C. L. (2020)., 29 Sept. 2020, https://doi.org/10.3389/fsufs.2020.534219.

Duff, Hannah, et al. "Precision Agroecology." *Sustainability* , vol. 14, no. 1, 23 Dec. 2021, p. 106, https://doi.org/10.3390/su14010106. Accessed 16 Jan. 2022.

Expertise, Sohrab Jam | Local SEO & Semantic SEO. "Community Engagement and Education for Recycling: Best Practices and Strategies." *Medium* , 8 Mar. 2024, medium.com/@sohrabjam/community-engagement-and-education-for-recycling-best-practices-and-strategies-b2b1485f1fce.

"Flooding: Adaptation Strategies: ERIT: Environmental Resilience Institute Part of the Prepared for Environmental Change Grand Challenge: Indiana University." *Environmental Resilience Institute Part of the Prepared for Environmental Change Grand Challenge* , eri.iu.edu/erit/strategies/flooding.html. Accessed 11 Aug. 2022.

Fox, Mary, et al. "Integrating Public Health into Climate Change Policy and Planning: State of Practice Update." *International Journal of Environmental Research and Public Health* , vol. 16, no. 18, 4 Sept. 2019, p. 3232, https://doi.org/10.3390/ijerph16183232.

Hoffman, Kira M., et al. "Conservation of Earth's Biodiversity Is Embedded in Indigenous Fire Stewardship." *Proceedings of the National Academy of Sciences* , vol. 118, no. 32, 10 Aug. 2021, www.pnas.org/content/118/32/e2105073118.short , https://doi.org/10.1073/pnas.2105073118.

Jeni, John. "Eco-Friendly Lifestyle Choices Gaining Popularity in the U.S." *Medium* , Medium, 17 Sept. 2024,

eliteautorepairflorida.medium.com/eco-
friendly-lifestyle-choices-gaining-popularity-
in-the-u-s-bc87bb167b92.

Knutson, Tom. "Global Warming and
Hurricanes – Geophysical Fluid Dynamics
Laboratory." *Noaa.gov* , 2021,
www.gfdl.noaa.gov/global-warming-and-
hurricanes/.

MacCarthy, James, et al. "New Data Confirms:
Forest Fires Are Getting Worse."
Www.wri.org , vol. (2023, August 29)., 29
Aug. 2023, www.wri.org/insights/global-
trends-forest-fires.

Mass Challenge. "Agriculture Innovation: 10
Tech Trends to Watch in 2022." *Mass
Challenge* , 26 Jan. 2022, masschallenge.org/
articles/agriculture-innovation/.

Mimura, Nobuo. "Sea-Level Rise Caused by
Climate Change and Its Implications for
Society." *Proceedings of the Japan Academy,
Series B* , vol. 89, no. 7, 25 July 2013, pp. 281–
301, www.ncbi.nlm.nih.gov/pmc/articles/
PMC3758961/ , https://doi.org/10.2183/pjab.
89.281.

Mojahed, Nooshin, et al. "Climate Crises and Developing Vector-Borne Diseases: A Narrative Review." *Iranian Journal of Public Health* , vol. 51, no. 12, 26 Dec. 2022, www.ncbi.nlm.nih.gov/pmc/articles/ PMC9874214/ , https://doi.org/10.18502/ ijph.v51i12.11457.

National Park Service. "Indigenous Fire Practices Shape Our Land." *Www.nps.gov* , 4 Feb. 2022, www.nps.gov/subjects/fire/ indigenous-fire-practices-shape-our-land.htm.

Native Peoples and the Environment Class Notes | Fiveable. (2024). "Restoration of Degraded Ecosystems Using Traditional Knowledge | Native Peoples and the Environment Class Notes | Fiveable." *Fiveable.me* , 2024, library.fiveable.me/native-people-their-environment/unit-10/restoration-degraded-ecosystems-traditional-knowledge/ study-guide/7LpV3jtQjr8Ts9Ip.

Padhy, Susanta Kumar, et al. "Mental Health Effects of Climate Change." *Indian Journal of Occupational and Environmental Medicine* , vol. 19, no. 1, 2015, p. 3, www.ncbi.nlm.nih.gov/pmc/articles/ PMC4446935/ , https://doi.org/ 10.4103/0019-5278.156997.

Pronthego.com. "PR on the GO Local Entrepreneurs and the UN's Sustainability Agenda – Strategies for Business Success." *Pronthego.com* , 2024, pronthego.com/pages/ blog/local-entrepreneurs-and-the-UN-sustainability-agenda-strategies-for-business-success.

Santini, Nadia S., and Yosune Miquelajauregui. "The Restoration of Degraded Lands by Local Communities and Indigenous Peoples." *Frontiers in Conservation Science* , vol. 3, no. Santini, N. S., & Miquelajauregui, Y. (2022), 25 Apr. 2022, https://doi.org/10.3389/fcosc. 2022.873659.

"Simulations Suggest Ice-Free Arctic Summers by 2050." *ESA Climate Office* , 13 May 2020, climate.esa.int/en/projects/sea-ice/news-and-events/news/simulations-suggest-ice-free-arctic-summers-2050/.

"Social Equity | U.S. Climate Resilience Toolkit." *Climate.gov* , 2016,

toolkit.climate.gov/topics/built-environment/
social-equity.

Steiner, Nadja S., et al. "Climate Change
Impacts on Sea-Ice Ecosystems and Associated
Ecosystem Services." *Elementa: Science of the
Anthropocene* , vol. 9, no. 1, 2021, https://
doi.org/10.1525/elementa.2021.00007.

"Sustainability in Mexico - Original Travel."
Www.originaltravel.co.uk   ,
www.originaltravel.co.uk/travel-guide/mexico/
sustainability.

Taylor, Matt. "Extreme Weather: How It Is
Connected to Climate Change?" *BBC News* , 9
Aug. 2021, www.bbc.com/news/science-
environment-58073295.

Team, S. E. O. "The Most Eco-Friendly
Transportation." *Roll'eat USA* , 17 Jan. 2024,
rolleatusa.com/eco-friendly-transportation/.

"Technology Could Boost Renewable Energy
Storage." *ScienceDaily* , 2024,
www.sciencedaily.com/releases/
2024/09/240916153438.htm.

"The Environmental Impact of Energy-Efficient
Homes." *Atlashomeenergy.com* , Atlas Home

Energy, 2024, www.atlashomeenergy.com/ blog/posts/view/the-environmental-impact-of-energy-efficient-homes.

Thomson, Madeleine C., and Lawrence R. Stanberry. "Climate Change and Vectorborne Diseases." *New England Journal of Medicine* , vol. 387, no. 21, 24 Nov. 2022, pp. 1969–1978, www.nejm.org/doi/full/10.1056/ NEJMra2200092 , https://doi.org/10.1056/ nejmra2200092.

"Top Strategies to Boost Your Home's Energy Efficiency · Greener Wisdom." *Greenerwisdom.com* , 2024, www.greenerwisdom.com/blog/homes-energy-efficiency.

UN Environment Programme. "Indigenous People and Nature: A Tradition of Conservation." *UN Environment* , 2017, www.unep.org/news-and-stories/story/ indigenous-people-and-nature-tradition-conservation.

Weiskopf, Sarah R. "Climate Change Effects on Biodiversity, Ecosystems, Ecosystem Services, and Natural Resource Management in the

United States." *Science of the Total Environment* , vol. 733, no. 733, 1 Sept. 2020, p. 137782, www.sciencedirect.com/science/article/pii/S0048969720312948 , https://doi.org/10.1016/j.scitotenv.2020.137782.

World Health Organization. "Climate Change." *World Health Organization* , 12 Oct. 2023, www.who.int/news-room/fact-sheets/detail/climate-change-and-health.

World Wildlife Fund. "Deforestation and Forest Degradation." *World Wildlife Fund* , World Wildlife Fund, 4 Dec. 2018, www.worldwildlife.org/threats/deforestation-and-forest-degradation.

Www.freshapproach.org. "Who We Are – Fresh Approach." *Www.freshapproach.org* , 2024, www.freshapproach.org/who-we-are/.

Www.who.int. "A Strategic Framework for Emergency Preparedness." *Www.who.int* , 2024, www.who.int/publications/i/item/a-strategic-framework-for-emergency-preparedness.

Yamanouchi, Takashi, and Kumiko Takata. "Rapid Change of the Arctic Climate System and Its Global Influences - Overview of GRENE Arctic Climate Change Research Project (2011–2016)." *Polar Science* , vol. 25, no. 2020, September, Sept. 2020, p. 100548, https://doi.org/10.1016/j.polar.2020.100548.

Zhao, Mengqi, and Jan Boll. "Adaptation of Water Resources Management under Climate Change." *Frontiers in Water* , vol. 4, 20 Oct. 2022, https://doi.org/10.3389/frwa.2022.983228.

12 Examples of Climate-Resilient City Solutions . (n.d.). State of Green. https://stateofgreen.com/en/news/12-examples-of-climate-resilient-city-solutions/

Lin, B. B., Ossola, A., Alberti, M., Andersson, E., Bai, X., Dobbs, C., Elmqvist, T., Evans, K. L., Frantzeskaki, N., Fuller, R. A., Gaston, K. J., Haase, D., Jim, C. Y., Konijnendijk, C., Nagendra, H., Niemelä, J., McPhearson, T., Moomaw, W. R., Parnell, S., & Pataki, D. (2021, July 1). *Integrating solutions to adapt cities for climate change* . The Lancet Planetary

Health. https://doi.org/10.1016/S2542-5196(21)00135-2

Nieuwenhuijsen, M. J. (2021, December). *New urban models for more sustainable, liveable and healthier cities post covid19; reducing air pollution, noise and heat island effects and increasing green space and physical activity* . Environment International. https://doi.org/10.1016/j.envint.2021.106850

Satterthwaite, D., Archer, D., Colenbrander, S., Dodman, D., Hardoy, J., Mitlin, D., & Patel, S. (2020, February). *Building Resilience to Climate Change in Informal Settlements* . One Earth. https://doi.org/10.1016/j.oneear.2020.02.002

The White House. (2023, November 9). *FACT SHEET: Fifth National Climate Assessment Details Impacts of Climate Change on Regions*

Across the United States | OSTP . The White House. https://www.whitehouse.gov/ostp/news-updates/2023/11/09/fact-sheet-fifth-national-climate-assessment-details-impacts-of-climate-change-on-regions-across-the-united-states/

University of Michigan. (2023). *U.S. Cities Factsheet* . Center for Sustainable Systems. https://css.umich.edu/publications/factsheets/built-environment/us-cities-factsheet

U. S. Global Change Research Program. (2023). *Fifth National Climate Assessment* . Nca2023.Globalchange.gov. https://nca2023.globalchange.gov/

silviapellegrino. (2023, June 16). *The cities adapting best to climate change - CityMonitor* . CityMonitor. https://www.citymonitor.ai/

analysis/cities-adapting-prepared-for-climate-change/

Made in United States
North Haven, CT
19 November 2024